The Universe

THE UNIVERSE

From Flat Earth to Black Holes—and Beyond

Isaac Asimov

THIRD EDITION, REVISED

WALKER AND COMPANY
NEW YORK

First published in the United States of America
in 1966 by the Walker Publishing Company,
Inc.

Published simultaneously in Canada by Beaver-
books, Limited, Don Mills, Ontario.

ISBN: 0-8027-0655-X

Library of Congress Catalog Card Number:
79-48052

Printed in the United States of America

To Fred L. Whipple and Carl Sagan
who know much more about it than I do

Table of Contents:

The Universe

1

The Earth

Introduction

In the last few years, astronomers have excited themselves and the public enormously by the discoveries they have been making in unimaginably distant outer space.

Phrases like quasars and pulsars are making headlines. Points of light billions of billions of miles away set scientists to wondering about the far past and far future of the Universe.

Does the Universe extend forever or is there an end somewhere? Does it expand and contract like an accordion, with each in and out motion lasting billions of years? Was there a time when it exploded once and for all and will the flying fragments separate until our own fragment is virtually alone in the universe? Does the Universe renew itself and is it eternal, unborn, and undying?

We are a fortunate generation, for we are watching a period of astronomy in which the answers to such questions and to many others equally intriguing may actually be at hand.

The situation is an unexpected one, too. The celestial objects that are opening new vistas for astronomers were not known before the 1960's. The rockets and satellites that are now feeding them so much data were not blasting off before the 1950's. The radio telescopes that disclosed unexpected wonders of the Universe were not in existence before the 1940's.

In fact, if we go back 2500 years to, say, 600 B.C., we will find that the entire Universe known to man was but a patch of flat ground, and not a very large patch either.

That is still all that man is directly aware of today; just a patch of flat ground and, of course, the sky overhead, with small luminous objects shining in it. Nor does the sky seem to be very far above our heads.

By what process of reasoning then, did the narrow surroundings visible to ourselves fade outward and outward and outward until no man's mind can possibly grasp the size of the Universe we now speak of, or imagine the tiny insignificance of our physical surroundings in comparison with it?

In this book I want to trace the steps by which man's grasp of the Universe as a whole ("cosmology") and of its origin and development ("cosmogony") widened and deepened.

The Flat Earth

In 600 B.C., the Assyrian Empire had just fallen. At its height, it had extended from Egypt to Babylonia, for an extreme length of 1400 miles. It was soon to be replaced by the Persian Empire, which extended from Cyrenaica to Kashmir, for an extreme length of 3000 miles.

Undoubtedly, the common folk of these empires had only the vaguest notion of the extent of the realm and were content to live and die on their own few acres or, on some occasions, to travel from village to neighboring village. Travelers and soldiers, however, must have had some concept of the vastness of these empires and of the still greater vastness of what must lie beyond.

There must have been in the ancient empires those who occupied themselves with what might be considered the first cosmological problem facing scholars: Is there an end to the Earth?

To be sure, no man in ancient times, however far he traveled, ever came to any actual end of the Earth. At most, he reached the shore of an ocean whose limits were beyond the horizon. If he transferred to a ship and sailed outward, he never succeeded in reaching the end either.

Did that mean there was no end?

The answer to that question depended on the general shape

one assumed for the Earth.

All men, before the time of the Greeks, made the assumption that the Earth was flat, as indeed it appears to be, barring the minor irregularities of the mountains and valleys. If any pre-Greek ancient thought otherwise, his name has not come down to us and the record of his thinking has not survived.

Yet if the Earth were indeed flat, an end of some sort would seem an almost foregone conclusion. The alternative would be a flat surface that would go on forever and forever — one that would be infinite in extent, in other words. This is a most uncomfortable concept; throughout history, men have tended to avoid the concept of endlessness in either space or time as something impossible to grasp and understand and therefore something that cannot easily be worked with or reasoned about.

On the other hand, if the Earth does have an end — if it is finite — there are other difficulties. Would not people fall off if they approached that end too closely?

Of course, it might be that the dry land was surrounded by ocean on all sides so that people could not approach the end unless they deliberately boarded a ship and sailed out of sight of land; far out of sight. As late as the time of Columbus, in fact, this was indeed a very real fear for many seamen.

The thought of such a watery protection of mankind raised another point, however. What was to prevent the ocean from pouring off the ends and draining away from the Earth?

One way out of this dilemma was to suppose that the sky above was a solid shield as, indeed, it appears to be[1], and that it came down to meet the Earth on all sides, as it appears to do. In that case, the entire Universe might be thought to consist of a kind of box, with the sky making up the curved top and sides, while the flat bottom is the sea and dry land on which man and all other things live and move.

What might the shape and size of such a "box-Universe" be?

To many, it seemed a rectangular slab. It is an interesting accident of history and geography that the first civilizations on the Nile, the Tigris-Euphrates, and the Indus Rivers were separated east and west, rather than north and south. Moreover, the Mediterranean Sea runs east and west. The dim geographical knowledge of early civilized man therefore expanded more easily

---

[1]   The biblical term "firmament" attests to the primitive belief of the sky as a "firm" object, a solid substance.

east and west than north and south. It seems reasonable to picture the "box-Universe" then as considerably longer east-west than north-south.

The Greeks, however, seemed to have a stronger sense of geometric proportion and symmetry. They tended to think of the Earth as a circular slab, with Greece, of course, in the center. This flat slab consisted chiefly of land, with a rim of water ("the Ocean River") from which the Mediterranean Sea extended inward to the center.

By 500 B.C., the first scientific geographer among the Greeks, Hecataeus of Miletus (birth and death dates unknown), considered this circular slab to have a diameter of perhaps 5000 miles at most. This would make the area of the flat Earth about 20,000,000 square miles. Such a figure would certainly have seemed ample, and even enormous, to the men of Hecataeus' time, but it represents only a tenth of the Earth's actual surface.

Then, too, what kept the box-Universe, whatever its size and shape, in place? In the vision of a flat Earth, which we are considering, "down" means one particular direction and all things that are heavy and earthy fall "downward." Why does the Earth itself not do so?

One might suppose that the material of which the flat Earth is composed, the land we stand on, simply continues downward forever. If so, we are faced once more with the concept of infinity. To avoid that, people might suppose instead that the Earth was standing on something. The Hindus placed it on four pillars, for instance.

But that only postponed the difficulty. On what were the four pillars standing? On elephants! And on what were the elephants standing? On a gigantic turtle! And the turtle? It swam in a gigantic ocean! And this ocean ——

In short, the assumption of a flat earth, however much it might seem to be "commonsensical," inevitably involved one in philosophic difficulties of the most serious sort.

The Spherical Earth

Indeed, the flat Earth did not even appear to be commonsensical, if one used one's eyes properly. If the Earth were really flat, then the same stars ought to be visible in the sky from all points (with some minor differences due to foreshortening, perhaps). Yet it was the universal experience of travelers that if one traveled north, some stars disappeared beyond the southern horizon and new stars appeared from behind the northern horizon. If one traveled south, the situation was reversed. This could most easily be explained by supposing that the Earth curved in a north-south direction. (Whether there was a similar east-west effect was obscured by the general east-west motion of the entire sky, which made one complete turn every twenty-four hours.)

The Greek philosopher Anaximander of Miletus (611–546 B.C.) suggested therefore that men lived on the surface of a cylinder that was curved north and south. He was the first man, as far as we now know, to suggest any shape for the Earth's surface other than flat, and the suggestion was perhaps made about 550 B.C.

Yet a cylindrical Earth was insufficient. It was the experience of men who lived on the seashore and dealt with ships that vessels heading out to sea did not merely grow smaller and smaller until they disappeared into an infinitesimal point, as would be expected if the Earth were flat. Instead, they disappeared while still perceptibly larger than points and did so hull-first as though they were moving beyond the top of a hill. This would be exactly what was to be expected if the surface of the Earth were curved. What is more, ships disappeared in much the same fashion no matter toward which point of the compass they moved. Therefore the Earth was curved not only north-south, but in all directions equally; and the only surface that curves in all directions equally is that of a sphere.

It also seemed to Greek astronomers that an eclipse of the Moon could best be explained by supposing that the Moon and Sun were on opposite sides of the Earth and that it was the Earth's

The disappearing hull

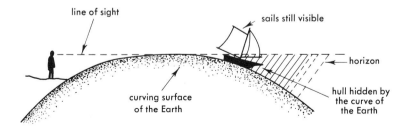

line of sight

sails still visible

horizon

curving surface of the Earth

hull hidden by the curve of the Earth

shadow (cast by the Sun) that fell on the Moon and eclipsed it. This shadow was always seen to be circular in cross section, no matter what position the Moon and Sun held with respect to the Earth. The only solid that casts a shadow with a circular cross section in all directions is a sphere.

Close observation, then, would show that the Earth's surface is not flat but spherical. It appears flat to the casual glance only because it is such a large sphere that the small portion of it visible to the eye has a curvature too gentle to detect.

As far as we now know, the first person to suggest that the Earth was a sphere was the Greek philosopher Philolaus of Tarentum (480– ? B.C.), who made the suggestion about 450 B.C.

The concept of a spherical Earth at once put to rest any problem of an "end" to the Earth, without introducing the concept of infinity. A sphere had a surface of finite size but one without an end; it was finite but unbounded.

About a century after Philolaus, the Greek philosopher Aristotle of Stagira (384–322 B.C.) summarized the implications of a spherical Earth.

One had to consider "down" not as a definite direction but as a relative one. If it were a definite direction as we sometimes think it is when we point to our feet, then the whole spherical Earth might be expected to fall downward forever — or until it comes to rest on something that is solid and has an infinite downward extension.

Suppose, however, we merely define "down" as the direction pointing to the center of the Earth. When we say things naturally "fall downward," we mean they naturally fall toward the center of the Earth. In that case, nothing falls off the Earth and the people on the other side of the globe from ourselves have no sensation of standing upside down.

The Earth itself cannot fall because every part of it has already fallen as much as possible and has reached as closely as possible to the Earth's center. Indeed, that is why the Earth must be a sphere, for the sphere has the property that the total distance of all parts of itself to its center is less than that of any solid of the same size but of a different shape.

By 350 B.C., then, we can say that it was firmly established that the Earth was a sphere. This concept has been accepted ever since by the educated of the Western world.

The concept was so satisfying and so free of paradox that men accepted it even in the absence of direct proof. It was not until 1522 A.D., eighteen centuries after Aristotle, when one surviving ship of an expedition originally led by the Portuguese navigator Ferdinand Magellan (1480–1521) sailed into port, that anyone actually succeeded in sailing completely around the Earth, thereby establishing direct proof that it was not flat.

Today we have established the sphericity of the Earth on an actual "seeing-is-believing" basis. Rockets climbed high enough in the late 1940's to take pictures of large sections of Earth's surface that show its visible spherical curvature.[2]

The Size of the Earth

With the Earth established as a sphere, the question of its size becomes more meaningful than ever. It would be difficult to establish the size of a finite flat Earth without actually pacing it off. A spherical Earth, however, produces effects that vary with the size of the sphere.

If the sphere of the Earth were extremely large, for instance, the effects produced by its sphericity would be too small to detect easily. The view of the stars would not change perceptibly as one traveled a mere few hundred miles north or south; ships would not disappear hull-first while they were still large enough to be seen; the cross-section of Earth's shadow against the Moon would seem straight, for its curve would be too gentle to detect.

The mere fact, then, that the effects of sphericity *were* noticeable meant not only that the Earth was a sphere but that it was one of rather limited size — large, perhaps, but not enormously large.

Still, how was one to measure that size with precision? Greek geographers could set a lower limit. By 250 B.C., they knew from experience that there was land to the west somewhat beyond the Strait of Gibraltar and other land to the east as far as India, an extreme distance of some 6000 miles (well above Hecataeus' apparently generous estimate of two and a half centuries earlier). Over that extent, Earth's surface clearly had not doubled back on itself, so the circumference of the Earth had to be greater than 6000 miles; but how much greater one could not say.

The first to suggest an answer based on observation was the Greek philosopher Eratosthenes of Cyrene (276–196 B.C.). He

[2]   To be sure, the Earth is not an *exact* sphere, but the departure from the exactly spherical is too small to be visible to the eye. Viewed from space, Earth would seem a perfect sphere.

knew, or was told, that on the summer solstice, June 21, when the noonday Sun was as near the zenith as it ever was, it was actually at the zenith at the city of Syene, Egypt (the modern Aswan). This could be demonstrated by the fact that a stick placed upright in the ground would then cast no shadow. At the same time, a stick placed upright in the ground in Alexandria, 500 miles north of Syene, would cast a short shadow, one that indicated the noonday Sun was a little over 7 degrees south of the zenith.

If the Earth were flat, the Sun would be virtually overhead both at Syene and at Alexandria simultaneously. The fact that it was overhead at one but not the other was proof, in itself, that the Earth's surface curved over the space between the cities. The stick in the ground at one city did not, so to speak, point in the same direction as did that at the other. While one pointed at the Sun, the other did not.

The greater the curvature, the greater the difference in direction in which the sticks pointed, and the greater the difference in the shadows. Eratosthenes carefully proved his calculations by geometry, but we can take the proof for granted and simply say that if a difference of a little over 7 degrees is 500 miles, then a difference of 360 degrees (the complete turn about a circle's circumference) must be about 25,000 miles to keep the proportion the same.

If the circumference of a sphere is known, then its diameter is known, too. The diameter is equal to the circumference divided by π ("pi"), a quantity equal to about 3.14. Eratosthenes therefore concluded that the Earth has a circumference of about 25,000 miles and a diameter of about 8000 miles.

The surface area of such a sphere is about 200,000,000 square miles, at least six times the largest area of the known world in ancient times. Eratosthenes' sphere was apparently a bit too large for the Greeks, and when later astronomers repeated his observations and obtained smaller figures (18,000 miles in circumference and 5700 miles in diameter for a surface area of 100,000,000 square miles), those smaller figures were accepted eagerly. The smaller figures prevailed throughout medieval times, and Columbus used them to prove that the westward journey from Spain to Asia was practical for ships of that time. In fact it was not; his voyage was a success only because the Americas occupied the space where he thought Asia would stand.

It was only in 1522 with the return of the single remaining

ship of Magellan's fleet that the true size of the Earth was established beyond question and Eratosthenes was vindicated.

The latest determinations make the circumference of the Earth at the Equator 24,902.4 miles. The diameter of the Earth varies slightly in different directions because the Earth is not an exact sphere, but its average length is 7917.78 miles. Its surface area is 196,950,000 square miles.

2

The Solar System

The Moon

If the Earth were all the Universe, the Greeks would have solved the essential problem of cosmology 2000 years ago. The Earth, however, is *not* all the Universe as the Greeks well knew. There is also the sky overhead.

As long as the Earth was assumed to be flat, it was perfectly possible to consider the sky to be a solid dome that closed down on that flat Earth on all sides. The enclosure it would form would not need to be terribly high, either. If it were ten miles high, for instance, that would suffice to enclose the highest mountains and the clouds.

If, however, the Earth were a sphere, the sky had to be a second, larger sphere enclosing it. It was the sphere of the sky (the "celestial sphere") that bounded the Universe, and it would therefore be of great interest to know its dimensions.

For all that could be told by casual observation, the celestial sphere might still hug the spherical Earth and be removed only ten miles from its surface in all directions. If the diameter of the Earth were 8000 miles, then that of the sky might be 8020 miles.

But let us not be satisfied with merely casual observation, for the Greeks — and the Babylonians and Egyptians before them — certainly were not.

The celestial sphere appears to revolve about the Earth once in twenty-four hours. In so doing, it seems to carry the stars with it "all in one piece." That is, the stars do not shift position relative to one another but remain fixed in place year after year and generation after generation (hence the "fixed stars"). It seemed natural to believe that the stars were attached to the vault of the sky like so many luminous pinheads, and until the seventeenth century that was, indeed, the common opinion.

However, even prehistoric man must have noticed that some of the heavenly bodies moved in relation to the stars and were near one star at one time and near a different star at another. These bodies, therefore, could not be attached to the vault of the sky but must be closer to the Earth than was the sky itself.

There were seven such bodies known to the ancients, and the names we know them by are, in order of brightness, the Sun,[1] the Moon, Venus, Jupiter, Mars, Saturn, and Mercury. These seven bodies were called "planetes" ("wanderers") by the Greeks, because they wandered among the stars. The word has come down to us as "planets."

It was possible to make a stab at judging which of the planets were closer and which farther. For instance, the Moon passed in front of the Sun at every Solar eclipse, so the Moon must be closer to the Earth's surface than the Sun is.

To make other distance judgments, the ancients relied on the relative speeds of the planetary motions among the stars. (We know from experience that the nearer a moving object is, the faster it seems to move. An airplane near the ground moves with frightening velocity, whereas the same airplane a mile high scarcely seems to move at all, although it might actually be speeding more quickly then than when it was near us.)

From the relative speeds of their motion against the stars, the Greeks judged the Moon to be the closest of the seven planets. The remainder, in order of increasing distance, were considered to be Mercury, Venus, Sun, Mars, Jupiter, and Saturn.

If, then, one wished to determine the distance of the heavenly bodies, it stands to reason one must start with the Moon. If the

[1]   The position of the Sun relative to the stars cannot be observed directly. It can, however, be observed that the stars visible at midnight differ somewhat from night to night. This is taken to mean that the Sun slowly moves among the stars, blanking out a slightly different part of the sky each day and leaving a slightly different part visible at night. It takes 365¼ days for the Sun to make a complete circuit of the sky in this fashion.

distance of the Moon cannot be determined, there is little chance of determining the distance of any other heavenly body.

The first to make a serious attempt to determine the distance of the Moon was the Greek astronomer Aristarchus of Samos (320–250 B.C.). He made use of observations during a Lunar eclipse. As the Earth's shadow fell across the Moon, one could tell from the curvatures of its edge how large its cross section was compared to the size of the Moon. If it is assumed that the Sun is much farther away than the Moon, then Aristarchus could use ordinary geometry to tell how far the Moon must be from the Earth in order to allow Earth's shadow to shrink down to the observed dimensions.

This method was improved and refined a little over a century later by another Greek astronomer, the greatest of ancient times, Hipparchus of Nicaea (190–120 B.C.).

Hipparchus came to the conclusion that the distance of the Moon from the Earth was equal to just about thirty times the diameter of the Earth. If Eratosthenes' figure for Earth's diameter — 8000 miles — is accepted, then the Moon's distance from the Earth is 240,000 miles.

This is an excellent figure considering the state of the art at the time. The best modern figure for the average distance of the Moon from the Earth, center to center, is 238,854.7 miles. This is only an average figure because the Moon does not move around the Earth in a perfect circle. It is a little closer to the Earth at some times than at others. The closest it ever gets to Earth ("perigee") is 221,463 miles and the farthest ("apogee") is 252,710 miles.

Knowing this distance, one can calculate the actual diameter of the Moon from its apparent size. It turns out to be 2160 miles through and to have a circumference, therefore, of 6800 miles. It is distinctly smaller than the Earth but is still of respectable size.

The distance of the Moon, once determined, disposed once and for all of any notion that the sky might be fairly close overhead. By Greek standards it was already seen to be an absolutely enormous distance. Even the closest heavenly body was a quarter of a million miles away and all the others had to be farther, perhaps much farther.

Could one probe further? What about the Sun?

Aristarchus realized that when the Moon was exactly at the first quarter (or last quarter), it, the Sun, and the Earth were at the vertices of a right triangle. By measuring the angle separating the Moon and the Sun as seen from the Earth, one could then use simple trigonometry to get the ratio of the distances of the Moon and the Sun. Then, if the distance of the Moon were determined, that of the Sun could be calculated.

Unfortunately for Aristarchus, measurements of angles in the heavens are rather difficult to obtain without good instruments, and determining the exact time of the Moon's first quarter is far from easy. The theory he worked with was mathematically perfect, but his measurements were a trifle off, enough to give him widely inaccurate results. He concluded that the Sun was twenty times as distant as the Moon. If the Moon is 240,000 miles from Earth, the Sun would then be a little less than 5,000,000 miles away, which is a gross underestimate (but which served as additional evidence for the unexpected largeness of the Universe).

By 150 B.C., then, we can say that in four centuries of careful astronomy, the Greeks had accurately determined the shape and dimensions of the Earth and the distance of the Moon, but had not managed to probe very far beyond that. They could conclude that the Universe was a huge sphere that was *at least* several million miles across, and at its center they placed an Earth-Moon system possessing dimensions we still accept today.

The Sun

For 1800 years after Hipparchus, man's understanding of the dimensions of the Universe proceeded no further. There seemed no way to determine the distance of any of the planets other than the Moon, and although a number of guesses were made as to the distance of the Sun, none was of any value.

One reason for this lack of progress after Hipparchus was that the Greeks had developed a model of the planetary system that was of only limited use. Hipparchus and those who followed considered the Earth to be the center of the Universe. The Moon and the other planets circled about the Earth (in a rather complicated fashion) and beyond them the vault of the stars also circled the Earth. This system was preserved in detail for posterity in the works of a late astronomer, Claudius Ptolemaeus, who lived in Egypt and wrote about 130 A.D. He is popularly known, in English, as Ptolemy; the "geocentric system" ("Earth-at-center") is often known as the "Ptolemaic system" in his honor.

Such a system made it possible for astronomers to calculate the apparent motions of the planets against the background of the stars with sufficient accuracy for the needs of the time. It did not, however, supply a model accurate enough to be of help in the determination of distances beyond the Moon.

The beginnings of a new model for the heavens was established by the Polish astronomer Nicolas Copernicus (1473–1543) who suggested, in a book published in 1543, on the very day of his death, that it was the Sun and not the Earth that was the center of the universe. The planetary system was, according to this view, truly a "Solar system," Sol being the Latin word for the Sun.

This had actually been suggested by Aristarchus nineteen centuries before, but it had then seemed a radical notion, too radical to accept. According to such a "heliocentric system" ("helios" is the Greek word for the Sun), the Earth would circle the Sun as the other planets did, and the whole vast mass of solid rock beneath man's feet would be flying through space without his being aware of it. There would then be six planets rather than seven: Mercury, Venus, Mars, Jupiter, Saturn, *and* Earth. The Sun would no longer be a planet but a motionless center. The Moon would not be a planet in the same sense as the rest, for it would be circling the Earth and not the Sun even in the heliocentric system. A body circling a planet was eventually named a "satellite" and that name describes the Moon.

The Copernican system began to make slow headway in the minds of astronomers, for by this time the geocentric view had been revealed to have many deficiencies. The mathematics involved in calculating planetary positions was cumbersome by the older system and yielded results that did not satisfy the painstaking observations of the new generations of astronomers of early modern times.

The heliocentric system yielded somewhat better results and simplified the mathematics as well. It was not an accurate model, however, for Copernicus still imagined planetary orbits to be combinations of perfect circles, which proved to be an inadequate point of view.

In 1609, an accurate model was finally advanced. The German astronomer Johannes Kepler (1571–1630) studied excellent observations of the position of the planet Mars, made by his old mentor, the Danish astronomer Tycho Brahe (1546–1601), and decided at last that the only geometric figure that would fit the observations would be an ellipse.[2] Kepler showed that the Sun

---

[2]   An ellipse is a kind of flattened circle which possesses two foci. If from any point on the ellipse, a straight line is drawn to each focus, the sum of the lengths of the two lines is always the same.

was at one focus of Mars' orbital ellipse.

The same eventually was shown to be true for all the planets revolving about the Sun, and for the Moon in its revolution about the Earth. The orbit was always an ellipse and the central body was always at one focus of that ellipse.

In 1619, Kepler went on to find that the average distances of the planets from the Sun were related in a simple mathematical fashion to their times of revolution about the Sun. It was easy to measure the times of revolution and, by comparing them, it was easy to calculate the relative distance of the different planets.

In short, a highly accurate model of the Solar system could now be drawn, with the orbits laid out in precise proportion. Unfortunately, however, by comparing times of revolution, you could only say that one planet might be twice as far from the Sun as another planet was, but you could not tell exactly how far either planet was from the Sun in miles. The model was there but not the scale on which it was constructed. Even this, however, gave a new idea as to the size of the Solar system, for it could be seen that the farthest planet known to the Greeks (or to Kepler), which was Saturn, was nearly ten times as far from the Sun as the Earth was.

Ellipse

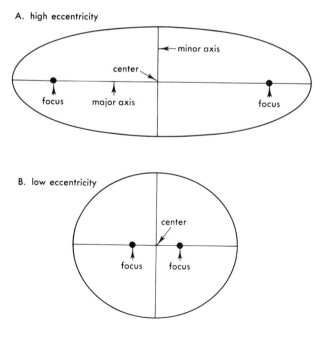

If the distance from the Earth to any one planet could be determined at any moment, it would give the scale. The distance of all the planets could then be determined. It remained only to determine one planetary distance correctly.

Parallax

To determine the distance of a planetary body, use could be made of a phenomenon known as parallax. This can be demonstrated most simply by holding one finger out in front of your eyes against some variegated background. Hold both finger and head steady and view the finger first with one eye and then with the other. You will see the finger shift position against the background as you switch from one eye to the other. If you bring the finger closer to your eye, it will shift position across a wider stretch of the background.

This happens because your two eyes are several inches apart, so that the line from the finger to one eye forms a perceptible angle with the line from the finger to the other eye. The two lines, extended toward the background, will show the finger to be appearing in two different positions. The closer the finger the greater the angle and therefore the greater the apparent shift. If your eyes were wider apart, that too would increase the angle formed by the two lines from your finger and create a larger shift of the finger against the background. (The background is usually so far away that lines from some point on it to both eyes make an angle too small to measure. The background can therefore be considered fixed.)

This same principle can be applied to a heavenly body. The Moon is so far away, of course, that switching eyes makes no difference. But suppose the Moon is viewed against the starry background of the heavens from two observatories that are hundreds of miles apart. The first observatory might see the Moon as having one edge a certain angular distance from a particular star; the other observatory might see the Moon at the same moment with that same edge a different angular distance from that particular star.

If the extent of the Moon's shift against the starry background (assuming the stars to be so far away as to remain fixed regardless of the change in position of the observatory) and the distance between the observatories are known, it is possible through the use of trigonometry to calculate the Moon's distance.

This could, in fact, be done, for the Moon's apparent shift against the stars with change in the position of the observer is quite large. Astronomers standardize this shift for the case in which one observer is viewing the Moon on the horizon while another is viewing it directly overhead. The base of the triangle is then equal to the radius of the Earth, and the angle at the Moon is the "equatorial horizontal parallax." Its value is observed to be 57.04 minutes of arc or 0.95 degrees of arc[3]. This shift is indeed sizable, for it is equal to twice the apparent diameter of the full Moon. It can be determined with sufficient accuracy, therefore, to give a good value for the Moon's distance. This distance, calculated by parallax, gave good agreement with the distance calculated by

[3] One degree (1°) is equal to 1/360 of the circumference of a circle. There are sixty minutes (60') in one degree, and sixty seconds (60'') in one minute.

Parallax

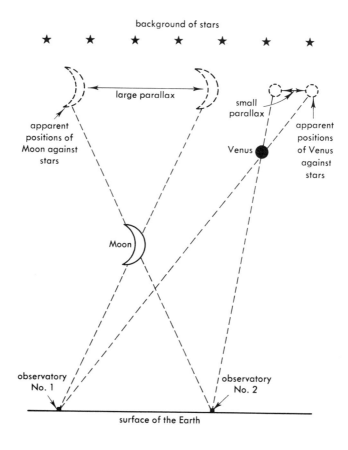

background of stars

large parallax

small parallax

apparent
positions of
Moon against
stars

apparent
positions
of Venus
against
stars

Venus

Moon

observatory
No. 1

observatory
No. 2

surface of the Earth

the earlier method that involved the Earth's shadow during a Lunar eclipse.

Unfortunately, the planets were so far distant that the change in position from observatory to observatory under conditions prevailing in 1600 produced a shift against the starry background that was too small to measure accurately.

Then, in 1608, came the invention (or reinvention) of the telescope by the Italian scientist Galileo Galilei (1564–1642). The telescope had the effect of magnifying the small shifts involved in parallax. A shift too small to detect with the naked eye became large enough to measure easily by telescope.

The closest planets (and therefore those with the largest parallaxes) are Venus and Mars. Venus, however, at its closest approach, is so close to the Sun in the sky that it cannot be observed (except in the rare cases when it can actually be seen against the Sun's disc, crossing it in "transit"). The logical target for parallax-determination beyond the Moon is, therefore, Mars.

In 1671 the first good telescopic determination of a planetary parallax was made. One of the observers was Jean Richer (1630-1696), a French astronomer who led a scientific expedition to Cayenne in French Guiana. The other was the Italian-French astronomer Giovanni Domenico Cassini (1625–1712), who stayed behind in Paris. Each observed Mars at as nearly the same time as possible and noted its position in relation to nearby stars. From the difference in position that was observed, and the known distance from Cayenne to Paris, the distance of Mars at that moment could be calculated.

Once that measurement was made, the scale of Kepler's model was known, and all other distances in the Solar system could be calculated. The distance of the Sun from the Earth, for instance, was calculated by Cassini to be 87,000,000 miles. This is about six million miles short of the actual figure, but it was excellent for a first attempt and may be considered the first useful determination of the dimensions of the Solar system.[4]

[4]   The Sun's parallax cannot be determined directly with any accuracy, for to do so would require an astronomer to work with one fixed point on the surface or edge of the globe. There are no fixed points on its blinding surface, and the tiny shift involved in its parallax (only about 8.8″ or 1/400 that of the Moon) cannot be made out. The planets, with much smaller and dimmer globes are more easily and accurately observed in some specific position, and a small parallax is much more easily detected. From that, the Sun's distance can be calculated much more satisfactorily than by any direct determination.

During the two centuries after Cassini's time, somewhat more accurate determinations of planetary parallaxes were made. Some of them involved Venus which, on occasion, passes directly between the Sun and the Earth and can be seen as a small, dark, circular body crossing the Sun's glowing disc. Such "transits" took place in 1761 and 1769, for instance. If the transit is carefully observed from different observatories, the time at which Venus first makes apparent contact with the Sun's disc, the time it leaves, and the duration of the transit all will be found to vary. From the amount by which they vary and from the distances between the observatories, the parallax of Venus can be calculated; from that, its distance; and from that, the distance of the Sun.

In 1835, the German astronomer Johann Franz Encke (1791–1865) used the data from Venus transits to deduce a distance of 95,370,000 miles for the Sun. This was a little too high but only by a little over two million miles.

The difficulty of getting still more accurate values lay in the fact that Mars and Venus did show up as tiny globes in the telescope and that this slightly blurred attempts at fixing the precise position of the planet. Venus was particularly disappointing because it had a thick atmosphere which produced optical effects that slightly obscured the exact moment of contact with the Sun's disc during a transit.

Then came an unexpected break. In 1801, the Italian astronomer Giuseppe Piazzi (1746–1826) found a small planet circling in an orbit between Mars and Jupiter. He called it Ceres; it proved to be somewhat less than 500 miles in diameter. As the century progressed, hundreds of even smaller planets were found, all circling between the orbits of Mars and Jupiter. These were the "asteroids." Then, in 1898, the German astronomer Karl Gustav Witt (1866–1946) discovered Eros, an asteroid which strayed out of the "asteroid belt." Part of its orbit passed within that of Mars and approached Earth's orbit rather closely.

In 1931, Eros was scheduled to approach Earth to a distance of only about two-thirds that of Venus, the nearest of the large planets. This close approach meant an unusually large and easily measured parallax. Furthermore, Eros is so small (only an estimated fifteen miles in its longest diameter) that it held no atmosphere to fuzz its outlines and, despite its closeness, remained a mere point of light. This meant that its position could be determined with a great deal of accuracy.

A vast international project was set up. Thousands of photographs were studied and eventually, from the parallax and position of Eros, it was determined that the Sun was just a bit less than 93,000,000 miles from the Earth. This is an average distance, since the Earth moves about the Sun in an ellipse and not a circle. At its closest approach to the Sun ("perihelion"), the Earth is at a distance of 91,400,000 miles from the Sun; at the farthest ("aphelion"), it is at a distance of 94,600,000 miles.

The Size of the Solar System

In recent years, something even better than parallax has turned up. Techniques have been developed whereby very short radio waves ("microwaves") of the type used in radar beams, can be sent out into space, bounced off a planet such as Venus, and the reflections received and detected. The microwaves move at a velocity that is accurately known, and the time lapse between emission and reception can be measured accurately. The distance traveled by the microwave beam on the round trip and, therefore, the distance of Venus at a given time, can be determined with greater precision than will be yielded by any parallax determination.

In 1961 such microwave reflections were received from Venus. Using the data thus obtained, the average distance of the Sun from the Earth was calculated to be 92,960,000 miles.

In 1961 such microwave reflections were received from Venus. Using the data obtained in such experiments, the average distance of the Sun from the Earth has been calculated to be 92,950,000 miles.

Making use of the Keplerian model, the distance of all the planets, either from the Earth at some particular time or from the Sun, can be calculated. It is more convenient to give the distance from the Sun since this changes less with time, and in a less complicated way, than does the distance from the Earth.

The distance can be expressed in four ways of particular interest.

First, it can be given in millions of miles. In the United States and Great Britain, miles are the common unit for measuring long distances.

Second, it can be given in millions of kilometers. The kilometer is the common unit for measuring long distances in civilized nations other than the Anglo-Saxon ones, and it is used by scientists everywhere, even in the United States and Great Britain. A kilometer is equal to 1093.6 yards or 0.62137 of a mile. It's length can be placed, with reasonable accuracy, at ⅝ of a mile.

Third, to avoid millions of miles or kilometers, the average distance of the Sun from the Earth may be set equal to an "astronomic unit" (abbreviated A.U.). Distances can then be given in A.U., where 1 A.U. is equal to 92,950,000 miles or 149,588,000 kilometers. It is reasonably accurate to say: 1 A.U. = 150,000,000 kilometers.

Fourth, the distance may be given in terms of the time it takes light (or any similar radiation, such as microwaves) to cross that distance. Light travels, in a vacuum, at a velocity of 299,792.5 kilometers per second, and this can be set at the even value of 300,000 kilometers per second without too much inaccuracy. This velocity is equivalent to 186,282 miles per second.

A distance of approximately 300,000 kilometers can therefore be set equal to "1 light-second" (the distance traveled by light in one second). Sixty times that, or 18,000,000 kilometers is "1 light-minute" and sixty times that or 1,080,000,000 kilometers is "1 light-hour." We are not too far off, if we think of a light-hour as equal to a billion kilometers.

With this in mind, let us consider the planets known to the ancients and prepare a list of their average distances from the Sun in each of these four units:

Planet	Average Distance from the Sun			
	Million miles	Million kilometers	Astronomic units	Light-hours
Mercury	35.9	57.9	0.387	0.0535
Venus	67.2	108.2	0.723	0.102
Earth	92.9	149.5	1.000	0.137
Mars	141.5	227.9	1.524	0.211
Jupiter	483.3	778.3	5.203	0.722
Saturn	886.1	1428.0	9.539	1.321

From the time of Cassini on, then, it was known that the diameter of the Solar system from one end of Saturn's orbit to the other, was nearly three billion kilometers (or two billion miles). The diameter of the imaginary sphere that included the planets known to the Greeks was not a matter of only millions of miles as the Greeks of Hipparchus' time had suspected, but was thousands of millions.

Even this turned out to be inadequate. The diameter of planetary orbits was doubled at a stroke in 1781, when the German-English astronomer William Herschel (1738–1822) discovered the planet Uranus. The diameter was doubled again, in two steps, when the French astronomer Urbain Jean Joseph Leverrier (1811–1877) discovered Neptune in 1846 and the American astronomer Clyde William Tombaugh (1906–   ) discovered Pluto in 1930.

The distances from the Sun of these outer members of the Solar System are given below:

	Average Distance from the Sun			
Planet	Million miles	Million kilometers	Astronomic units	Light-hours
Uranus	1782	2872	19.182	2.66
Neptune	2792	4498	30.058	4.26
Pluto	3671	5910	39.518	5.47

If we consider the orbit of Pluto, rather than that of Saturn, we see that the diameter of the Solar system is not three billion kilometers but twelve billion. A ray of light, which can travel a distance equal to the circumference of the Earth in 1/7 of a second and pass from the Earth to the Moon in 1¼ seconds, would take nearly a half a day to span the Solar system. The sky has indeed receded unutterably since Greek times.

Size of the Solar System

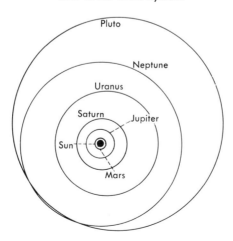

In fact, there is every reason to believe that Pluto does not mark the boundary of the Sun's dominion. Nor does this mean we must postulate still further undiscovered planets (although it is quite possible that small, very distant ones may exist). There are known bodies, which are easily seen on occasion, that undoubtedly recede to distances greater than that of Pluto at its farthest.

This fact was known even before the discovery of Uranus extended the boundaries of the strictly planetary portion of the Solar system. In 1684, the English scientist Isaac Newton (1642–1727) had worked out the law of universal gravitation. This law explained the existence of Kepler's model of the Solar system in a straightforward mathematical manner and made it possible to calculate the orbit of a body revolving about the Sun even when that body was only visible through part of its orbit.

This, in turn, made it possible to deal with comets, fuzzily luminous bodies that appeared now and then in the heavens. Through ancient and medieval times, astronomers had thought that comets arrived at irregular intervals in motions bound by no natural law—and the population in general was sure that their only purpose was to foretell disaster.

Newton's younger friend, the English astronomer Edmund Halley (1656–1742) tried, however, to apply gravitational calculations to the comets. He noted that certain spectacular comets seemed to show up in the skies at intervals of seventy-five or seventy-six years. He made the assumption, in 1704, that these comets were actually a single object moving in a regular orbit about the

Halley's comet

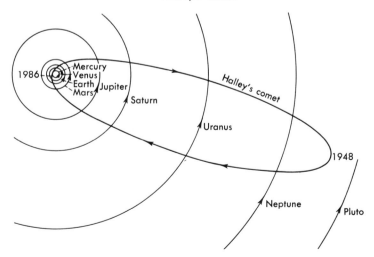

Sun—an orbit so elliptical that most of it was enormously far from the Earth. When the comet was distant from the Earth it was not visible, but every seventy-five or seventy-six years it was in the portion of its orbit near the Sun (and the Earth) and was then visible.

Halley calculated the orbit and predicted that the comet would once more be visible in 1758. The comet did indeed appear then (sixteen years after Halley's death) and it has been called "Halley's Comet" ever since. A search of historical records now shows that 28 different returns of this comet have been described, the earliest having been in 240 B.C.

At its closest approach to the Sun, Halley's Comet is only about ninety-million kilometers from the Sun, so that it moves slightly inside the orbit of Venus. At its farthest recession from the Sun, however, Halley's Comet moves out to a distance some 3½ times that of Saturn. At its aphelion, it is 5300 million kilometers from the Sun, which means that it recedes to a point well beyond Neptune's orbit. By 1760, then, the astronomers were perfectly aware that the Solar system was much larger than the Greeks had imagined and did not need the discovery of additional planets to tell them so.

Indeed, Halley's Comet is one of those that are relatively close to the Sun. There are some comets that move about the Sun in such fantastically elongated orbits that they return only after intervals of many centuries or even millennia. They recede from the Sun to distances not of mere billions of kilometers but, in all likelihood, hundreds of billions of kilometers. According to a suggestion made in 1950 by the Dutch astronomer Jan Hendrik Oort (1900–   ), there may even be a vast cloud of comets that remain at such immense distances throughout their orbits, and are therefore never seen.

It follows, then, that the Solar system may well have an extreme diameter of something like a thousand billion kilometers; that is, a trillion (1,000,000,000,000) kilometers or even more. It would take a ray of light forty *days* to span this distance. The diameter of the Solar system can thus be viewed as more than 1 "light-month."

Nor is the relative insignificance of the Earth a matter of distances only. The four outer planets: Jupiter, Saturn, Uranus, and Neptune expand into globes of measurable size when viewed through the telescope. Once the distance of these bodies came to be known, the apparent size of the globes could be turned into absolute measurements. Each of these outer planets turns out to be a giant compared with the Earth. And the size of the Sun makes it a giant compared with even the largest of the planets.

Object	Equatorial Diameter		
	Miles	Kilometers	Earth's diameter = .1
Earth	7927	12,753	1.00
Neptune	27,700	44,600	3.50
Uranus	29,200	47,000	3.68
Saturn	75,100	121,000	9.5
Jupiter	88,700	143,000	11.2
Sun	864,000	1,392,000	109.0

Then, too, each of the giant planets has a satellite system that dwarfs the Earth's. The first of the outer satellites to be discovered were the four largest of Jupiter, which were seen in 1610 by Galileo through his first primitive telescope. The last large satellite to be discovered was Neptune's satellite Triton, detected in 1846 by the English astronomer William Lassell (1799–1880). Additional small satellites were discovered; a second satellite of Neptune, Nereid, was noted as late as 1949 by the Dutch-American astronomer Gerard Peter Kuiper (1905–1973). The total number of known satellites in the Solar system, including our Moon, is now thirty-two.

The latest satellite to be discovered is one of Saturn's. The newcomer is so close to Saturn as to be ordinarily masked by light from that planet's ring system. In December 1966, however, the relative positions of Earth and Saturn were such that the rings were seen edgewise and therefore cast no light. The new satellite was seen for the first time and was named Janus.

A notion of the size of some of the satellite systems compared with that of the Earth is given in the following table:

Planet	Number of Satellites	Farthest Satellite	Average Distance of Farthest Satellite from its Planet	
			Miles	Kilometers
Earth	1	Moon	238,900	385,000
Uranus	5	Oberon	.368,000	591,500
Neptune	2	Nereid	3,461,000	5,540,000
Saturn	9	Phoebe	8,053,000	12,905,000
Jupiter	12	Hades*	14,700,000	23,600,000

* Unofficial name

CHAPTER 3

The Stars

The Vault of the Sky

If the Solar system were all there was to the Universe, the essence of the problem of its size would have been solved by 1700. However, the Solar system is *not* all there is to the Universe. There remain the stars.

In 1700, it was still possible to believe, that there was a solid vault bounding the Universe, one that contained the stars as luminous dots, and that this solid vault lay (possibly) not far outside the confines of the Solar system. Kepler's views on the matter were something like that, for instance.

Parallax measurements of the sort that had revealed the scale of the Solar system in the seventeenth century, were useless in connection with the stars and did nothing to disturb this "solid-sky" view. The separation of two neighboring stars never varied measurably no matter which observatories on Earth's surface made the measurements. A shift in base over the full width of the Earth caused no detectable change in the position of any star. This is not surprising, since the stars, even if barely outside the orbit of Saturn, would be too distant to show a parallax large enough to be measured in 1700.

The Earth's surface was not, however, the final resource of the astronomer in this respect. The Earth's diameter was not quite

8000 miles, but the entire globe moved through space as it pursued its revolution about the Sun. One side of that orbit was 186,000,000 miles from the other side. If, then, the positions of the stars were charted on some particular evening and then again on a particular evening a half-year later, the astronomer would be viewing the stars from two positions separated by 23,600 times as great a distance as that represented by the full width of the Earth's diameter. The expected parallax would be increased by that same ratio. Indeed, the position of a particular star might shift slightly each night as the Earth moved and, in the course of the year, the star would mark out a tiny ellipse in the sky—a kind of image of the Earth's orbit. The angular distance from the edge of the ellipse to its center would be the "stellar parallax."

This device cannot be used for the planets because in the course of the year, each planet marks out a rather complicated path of its own across the sky, one that masks any parallactic shifts caused by the Earth's motion. Trying to sort out the planet's own motion from that imposed on it by the Earth's would be extremely complicated and would yield results that were not as accurate as those arrived at by ordinary parallaxes. The stars, however, remain virtually fixed in place over the course of the year so that a useful parallactic shift might, conceivably, be obtained.

Such a shift, however, was *not* obtained. Indeed, the 1800's dawned without astronomers having been able to detect the stellar parallax of a single star.

Several reasons might be advanced for this.

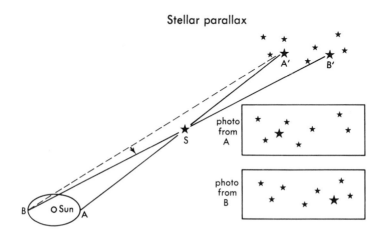

Stellar parallax

It might be, of course, that Copernicus and Kepler were wrong after all and that the Earth did not move around the Sun, but was the motionless center of the Universe. If so, no parallax over the course of a year would be expected. Indeed, when Copernicus first advanced the heliocentric theory, the absence of stellar parallaxes was used by his opponents as a strong argument against him. However, there were too many other reasons for accepting the heliocentric theory and at last it became firmly embedded in astronomic thought, despite the absence of stellar parallax. The Earth *does* move and the absence of parallax must be explained in other ways.

Despite the Earth's motion, parallaxes could still be absent if all the stars were virtually at the same distance. A parallax is visible only when the position of some relatively near object is sighted against some relatively distant one. If there were really a solid vault of the sky, the stars would all tend to shift in identical fashion with a comparatively small change in the position of the observer (and the shift of Earth in its orbital positions is small compared with even the small Universe some men accepted in 1700). There would then be no visible parallax.

But can the theory of the solid vault of the sky really be accepted? Arguments could be advanced for believing that stars might be at varying distances from us, even at widely varying distances. They might be distributed through a broad expanse of space, and there might be no solid boundary to the Universe at all.

For one thing, the stars vary in brightness. This is obvious to anyone who looks at the night sky. Hipparchus had been the first to attempt to reduce this difference in brightness to some sort of system. He divided the stars into six classes or "magnitudes." The brightest stars he considered to be first magnitude, the next brightest second magnitude, and so on down to the sixth magnitude which represents the dimmest stars that can be made out by the naked eye.

Modern astronomers measure the brightness of individual stars with instruments unavailable to the ancients and have defined the magnitudes with mathematical precision. A difference of five magnitudes (say between 1 and 6) represents a ratio, in brightness, of 100. In other words a star of magnitude 1 is 100 times as bright as one of magnitude 6. A difference of a single magnitude in brightness represents a ratio of 2.512, therefore, since $2.512 \times 2.512 \times 2.512 \times 2.512 \times 2.512$ is equal to 100.

Accurate measurements of brightness make it possible to define the magnitude of a star to fractions and even to a tenth of a magnitude. Thus, the bright star Aldebaran has a magnitude of 1.1, while Regulus, slightly dimmer, has a magnitude of 1.3. The still dimmer Polaris (the North Star) has a magnitude of 2.1, while Electra, one of the stars of the Pleiades, has a magnitude of 3.8.

There are a number of stars brighter than Aldebaran that have magnitudes less than 1.0. Procyon has a magnitude of 0.5 and the still brighter Vega one of 0.1. The very brightest stars in the heavens must be assigned negative magnitudes. Canopus has a magnitude of −0.7 and Sirius one of −1.4.

It is even possible to fit the planets, the Moon, and the Sun into this system. Venus, Mars, and Jupiter are all brighter at times than even the brightest star. Jupiter can attain a magnitude of −2.5; Mars a magnitude of −2.8; and Venus a magnitude of −4.3. The full Moon has a magnitude of −12.6 and the Sun one of −26.9.

Working in the other direction, stars dimmer than the sixth magnitude exist, too, although they are invisible to the naked eye. Galileo, when he first turned his telescope to the heavens in 1609, found hundreds of stars he could not see before. Stars of magnitudes 7, 8, 9, and so on, up the scale of numbers and down the scale of brightness, are recorded and studied. Our largest telescopes can make out myriads of stars with magnitudes more than 23.5.

If all stars were of the same intrinsic brightness (or "luminosity"), we could suppose that the difference in apparent brightness was entirely a matter of distance. Closer stars look brighter than distant ones, just as closer lamp posts seem more brilliantly lit than distant ones.

There was no reason in 1700 to assume that the stars were all of the same intrinsic brightness. It might just as well be that the stars were all at the same distance from Earth and that the difference in brightness was a real one; bright stars just happened to be more luminous than dim ones, as some light bulbs are intrinsically more luminous than others.

Another factor, however, dented the equal-distance hypothesis with considerably greater effectiveness.

The ancient Greeks had recorded the relative positions of the visible stars. The first to do so, in the 3rd Century B.C., were Aristyllos and Timochares of Alexandria. More systematic was Hipparchus who, about 134 B.C., had recorded the positions of over 800 stars. His was the first important "star map," and it was preserved for posterity by Ptolemy, who increased the number of stars it contained to more than a thousand.

In 1718, Halley, studying the positions of the stars, noted that at least three stars, Sirius, Procyon, and Arcturus, were not in the spots recorded by the Greeks. The difference in position was so great that it was unlikely that either the Greeks or Halley could have made a mistake. Halley found Arcturus, for instance, to be a full degree (twice the apparent width of the full Moon) away from the position recorded by the Greeks.

It seemed clear to Halley that these stars had moved. They were not truly fixed stars after all but had "proper motions" of their own. The proper motions of the stars were exceedingly slow compared with those of the planets and did not make themselves evident from day to day or even from year to year. But from generation to generation, the slow proper motion of the stars succeeds in displacing them perceptibly against the sky.

The mere existence of proper motions among the stars was a terrible blow to the solid sky hypothesis. At least some of the stars were not attached to the vault, and the feeling grew at once that none of the stars were; that, indeed, there was no vault.

Nevertheless, although the stars were not actually attached to any solid object, it was possible that they were all at virtually the same distance. There might still be a rather narrow shell of space through which stars might be distributed without being attached to anything.

This possibility was rendered unlikely by the fact that only a small minority of stars were found to display measurable proper motions. To be sure, a star might move without its motion being apparent to us, even over long periods, if it moved in a direction parallel to the line of sight. Yet if stars moved in any direction, at random, then at least as many ought to move more or less at right angles to the line of sight as parallel to it. In that case, if any stars at all showed a measureable proper motion, at least half of them ought to. Still, the keenest search revealed measurable proper motion to be very much the exception.

But what if we abandon the assumption that all stars are roughly at the same distance? Let us suppose, instead, that they are at widely varying distances. If all of them are moving at the same velocity, or over some reasonably narrow range of velocities, and in any direction at random, we can come to certain conclusions.

Of those stars moving more or less parallel to the line of sight, none will show measurable proper motions, whether near or far. Of those moving more or less at right angles to the line of sight, those that are nearer will have larger proper motions than those that are farther.

This association of comparatively large proper motion with comparative closeness is borne out by the fact that it is the bright stars that are most likely to show such proper motion. The first three stars found to possess proper motion—Sirius, Procyon, and Arcturus—are among the eight brightest stars in the sky. It is clear that a nearby star is apt to appear bright as well as to possess a proper motion. According to this view, it makes good sense to find a measurable proper motion only among very few stars. It may well be that only the nearest stars are close enough to show even a tiny proper motion and that beyond them lie vast myriads of stars too distant to show any perceptible motion at all, even over many centuries.

By the mid-eighteenth century, then, it was perfectly clear that there was neither a solid sky nor even a relatively narrow shell

Proper motion and distance

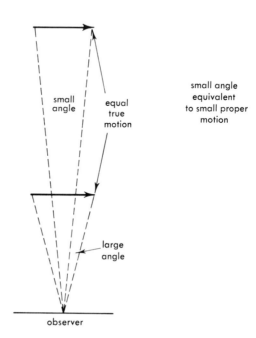

of stars. Rather the stars were widely spread out through an indefinitely vast expanse of space. This view had, in fact, been suggested by a few medieval scholars such as the German philosopher Nicholas of Cusa (1401–1464), but what had then been speculation had now become deduction from careful observation.

A Multiplicity of Suns

But then, if the Earth is moving, and if the stars are distributed at all distances, why do not the nearer ones show a stellar parallax with reference to the farther ones?

One explanation of this was so forcefully obvious that it received almost instant acceptance. Even the nearest stars were so distant that their parallaxes were too small to measure with the instruments of 1800. (Indeed, Copernicus had maintained just this in answer to those critics who held up the absence of stellar parallax as evidence against the revolution of the Earth about the Sun.)

In fact, it was possible to make rough estimates of the distances of the nearer stars by several logical lines of argument.

Suppose, for instance, we assume that the stars moved, in reality, as rapidly as do the planets. If so, we can estimate the distance at which such a motion is reduced in appearance to the tiny, almost immeasurably small crawl represented by a star's proper motion.

Thus, the star, with the greatest of all proper motions is "Barnard's star" (so named because it was discovered in 1916 by the American astronomer Edward Emerson Barnard [1857–1923]). Its proper motion is 10.3 seconds of arc per year.

To understand how little this is, remember that a circle is divided into 360 degrees, each degree into 60 minutes, and each minute into 60 seconds. A single second of arc represents 1/1,296,000 of the circuit of the sky. Since the Moon is 31 minutes of arc in diameter, a single second represents 1/1860 the diameter of the Moon. Jupiter looks like a mere point of light to us, but the diameter of its globe as seen by the naked eye is from 30 to 50 seconds of arc, depending on how close to us it is.

When we say, then, that Barnard's star moves 10.3 seconds of arc a year, we mean that it moves about 1/180 the diameter of the Moon, or only about a quarter the diameter of the Jupiter pinpoint in one year. And yet so comparatively large a proper motion is this that the star is sometimes called "Barnard's runaway star." It is much more usual for proper motions to be 1 second of arc per year or less.

Suppose, then, that Barnard's star actually moved as quickly across our line of sight as does Earth in its journey about the Sun — 18.5 miles per second. In one year, it would travel 585,000,000 miles. For such a length to mark off only 10.3 seconds of arc, Barnard's star would have to be something like ten trillion miles from us — thousands of times as far away from the Sun as Pluto is. And if this were so, the parallax Barnard's star would display would be only about 1 second of arc. If Barnard's star moved more quickly than does the Earth (as, in actual fact, it does), then it would be still more distant and would have a correspondingly smaller parallax.

If a star makes a small ellipse in the sky, one that is about 1 second of arc across or less, astronomers would be hard put to detect the fact. The size of the ellipse would be roughly that of a twenty-five cent piece viewed from a distance of four miles.

To be sure, proper motions of 1 second of arc per year can be observed without too much trouble, for proper motions go constantly in the same direction and pile up year by year. After a century, a star moving at 1 second of arc per year will have crawled nearly 2 minutes of arc across the sky and that shift is easy to observe by telescope. A parallactic motion, on the other hand, is eternally back and forth and does not pile up with the years.

If the stars are, in fact, tens of trillions of miles away at the very least, it becomes a matter of interest that we can see them at all. Even as supernally bright an object as the Sun would be seen as only a tiny point of light from a distance of ten trillion miles. It would appear to be merely a star. Conversely, any star seen from the distance of our Sun would increase so enormously in brightness as to appear another sunlike object itself.

We must, therefore, look on the Sun as a star which differs from all the other stars chiefly in that we see it from a distance of millions of miles, rather than trillions of miles as in every other case. We must also look on the Universe as a vast assemblage of suns of which our own is merely one.

Suppose, then, that the star Sirius were actually as luminous as the Sun and that its lesser light is only the result of its enormous distance from us. Sirius has a magnitude of — 1.6 and the Sun a magnitude of — 26.9. The Sun is 25.3 magnitudes brighter and each magnitude represents an increase of brightness by a ratio of 2.512. This means that the observed brightness of the Sun is 13,200,000,000 times that of Sirius.

The brightness of a light source varies inversely as the square of the distance. That is, if a light source is placed at two times its former distance, its brightness is reduced to $(\frac{1}{2})^2$ or $\frac{1}{4}$; if it is placed at five times its former distance, its brightness is reduced to $(\frac{1}{5})^2$ or $1/25$.

For Sirius to shine with only $1/13,200,000,000$ times the brightness of the Sun, it must be 115,000 times as far as the Sun, since $115,000 \times 115,000$ is equal to $13,200,000,000$. We know the Sun to be 93,000,000 miles from us and by this line of reasoning, Sirius ought to be a little over ten trillion miles away. In short, whether we argue from the proper motion of stars or from their brightness, we end with the same colossal distance and are forced to expect the same heartbreakingly small parallax.

To express a distance as trillions of miles is meaningless. Instead we can turn to units of distance as measured by the length traveled by light in a particular unit of time. I have already said that a light-hour is equal to 1,080,000,000 kilometers, which comes to about 670,000,000 miles. Suppose we consider the "light-year" instead.

At 300,000 kilometers per second, light will travel 9,440,-000,000,000 kilometers (or 5,880,000,000,000 miles) in one year. We can say, then, with rough accuracy, that a light-year is equal to ten trillion kilometers or six trillion miles.

Making use of this unit, we can see that Sirius (by the line of argument we have used above) is ten trillion miles, or nearly 2 light years, from us. Since Sirius is certainly one of the closer stars, whether one uses the criterion of brightness or of proper motion, one must conclude that stellar distances must all, at the very minimum, be measured in light-years.

Consider this in comparison to the Solar system. A ray of light that travels from the Sun to the Earth in 8 minutes and from the Sun to distant Pluto in 5½ hours could reach even the nearest stars only after a journey of years!

The Search for Stellar Parallax

As telescopes grew larger and better with the passing decades, hopes grew that the tiny stellar parallaxes might be detected and that the distances of at least some of the nearer stars might actually be determined directly rather than merely deduced from more or less rickety assumptions. Unfortunately, the closer stars were observed, the more complex the problem seemed to grow.

For instance, the close stars, which might have the largest parallaxes, were also those most likely to display considerable proper motion. This meant that a parallactic shift would have to be disentangled from the proper motion.

The situation became more complex in 1725 when the English astronomer James Bradley (1693–1762), in the course of careful determination of stellar positions, noted small shifts that did indeed cause a certain star to mark out a tiny ellipse in the sky in the course of the year. The trouble was that the star did not change position in accordance with what one would expect of a parallactic shift. Although the star should have appeared farthest south in December, it reached that point in March instead. The three-month delay continued all round the ellipse.

By 1728, Bradley had been able to show that this was the result of Earth's motion into the beams of light falling on it from the stars.

The analogy is usually made to a shower of rain. If the rain is falling vertically and a man with an umbrella is standing still, he need only hold the umbrella directly overhead. If, however, the man walks forward, he will move into some raindrops   that

Aberration

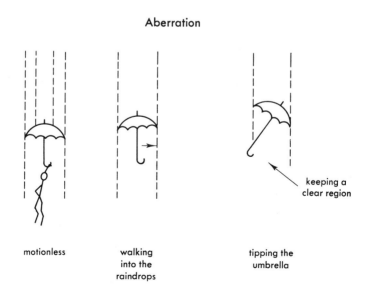

motionless   walking   tipping the
                  into the   umbrella
                  raindrops

keeping a
clear region

The Moon. *(Photograph from the Mount Wilson and Palomar Observatories.)*

Head of Halley's Comet, May 8, 1910. *(Photograph from the Mount Wilson and Palomar Observatories.)*

Saturn and Ring System. *(Photograph from the Mount Wilson and Palomar Observatories.)*

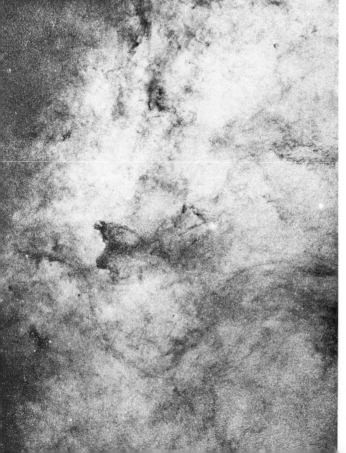

The Milky Way in Sagittarius. *(Photograph from the Mount Wilson and Palomar Observatory*

The Pleiades. *(Photograph from the Mount Wilson and Palomar Observatories.)*

The Great Hercules Cluster. *(Photograph from the Mount Wilson and Palomar Observatories.)*

The Large Magellanic
Cloud. (*Lick Observ-
atory Photograph.*)

Dark Nebulae. (*Photograph from the Mount Wilson and Palomar Observatories.*)

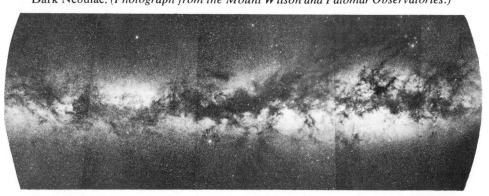

have just cleared his umbrella. He must, therefore, angle the umbrella forward slightly. The faster he moves the farther forward he must angle his umbrella, and if he changes direction, he must change the direction of the umbrella.

In the same way, since the Earth is moving through a "rain" of light-rays, the astronomer must angle his telescope very slightly by an amount that depends on the velocity of the Earth compared with the velocity of light. As the Earth changes its direction in revolving about the Sun, so the astronomer must change the direction of his telescope. In the end the star marks out an ellipse, but it is not a parallactic one.

This phenomenon is called the "aberration of light," and its effect on stellar position is larger than would be produced by parallax. A star can be displaced over a distance of as much as 40 seconds of arc by aberration, and if parallax is to be detected, it must be disentangled from this considerably larger effect.[1]

Bradley also discovered that the direction of the axis of the Earth relative to the stars moves slightly back and forth in a period of 18.6 years, as though the Earth were nodding. The motion is called "nutation" (from the Latin word meaning "nodding"). This motion is reflected in slight changes in the apparent position of the stars, and this, too, must be disentangled from any parallactic shift that might exist.

The search for stellar parallax, which led Bradley to the phenomenon of aberration, lured Herschel (the discoverer of Uranus) into a still more exciting discovery.

Herschel thought he might detect the tiny changes produced by stellar parallax with greater ease if he chose two stars that were extremely close together. The assumption was that such stars just happened to lie very close to the same line of sight but that one might be enormously more distant than the other. In that case, the nearer star would seem to move with respect to the

---

[1]   Although Bradley was trying to detect parallax when he detected aberration, this certainly cannot be considered a failure. The existence of the aberration of light is as good a piece of evidence for the motion of the Earth about the Sun as the detection of stellar parallax would have been. If the Earth were really a motionless body at the center of the Universe, there would be neither parallax nor aberration. By the eighteenth century, the geocentric theory scarcely needed another coffin nail, but Bradley's discovery provided one. In addition, the discovery of aberration allowed Bradley to estimate the velocity of light with better accuracy than had been possible previously.

other. (The suggestion that this be done had first been made by Galileo.)

Herschel found shifts almost at once, but the shifts could not be due to stellar parallax. A parallactic shift (after you subtract the effects of proper motion, aberration of light, and nutation) would have to produce a closed ellipse in the space of one year, but the shifts Herschel observed did not do this. Herschel did find his stars to be marking out ellipses, but they did so over an interval of time much longer than a year.

By 1793, he was convinced that what he was observing was the case of two stars circling each other about a common center of gravity. The shifts he saw depended only on the gravitational attraction of one star for the other and had nothing to do with parallax. The two stars were not independent entities that seemed close together only because both happened to be near the same line of sight (although this is certainly possible in some cases), but they were stars that were close together in actual fact as well as in appearance. Herschel had discovered "binary stars." Before

The motion of Sirius

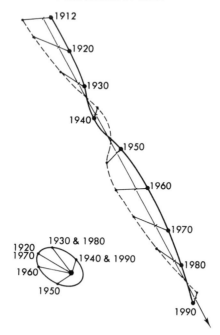

the end of his career, he had discovered 800 such binaries and since his time the number of known binaries has been increased to 60,000.

Indeed, the motions produced by virtue of the existence of a close neighbor star could be detected even when the neighbor star itself could not. In 1844, the German astronomer Friedrich Wilhelm Bessel (1784–1846), who was studying the star Sirius, found that its proper motion was not the straight line it ought to be. It was a wavy line. Had there been a companion star, it would have been easy to account for the wavy line as an elliptical motion about the companion, superimposed on the straight line of the proper motion. However, no companion star was visible. Bessel suggested that there was a "dark companion," a burnt-out ember, perhaps, of a once-bright star; a companion that could no longer be seen but that still existed and made its gravitational effect felt.

Bessel's suggestion was borne out in 1862. In that year the American astronomer Alvan Graham Clark (1832–1897) detected a dim spot of light near Sirius. This turned out to be Bessel's "dark companion." It was not completely dark, after all, but had a magnitude of 7.1.

The Distance of the Nearer Stars

As astronomic instruments continued to improve, and as the possible roadblocks in the form of complicating motions were better and better understood, hope continued to grow that stellar parallax might yet be detected. Attempts to achieve this goal grew more determined and sophisticated, and in the 1830's three separate attacks were made on the problem.

In South Africa, the Scottish astronomer Thomas Henderson (1798–1844) was carefully plotting the position of Alpha Centauri. This is the third brightest star in the sky (but is too far south to be seen from North Temperate latitudes) and therefore, he hoped, one of the closer ones.

In the Baltic provinces of Russia, the German-Russian astronomer Friedrich Georg Wilhelm von Struve (1793-1864) was plotting the position of Vega. This is the fourth brightest star, and it, too, might be one of the close ones.

In Königsberg in East Prussia, Bessel (who in the next decade would discover Sirius' companion) was taking another tack. It was not brightness he was using as a criterion for closeness, but rapid proper motion. For the purpose, he selected a star known as 61 Cygni (in the constellation Cygnus, the Swan) which had been found to have a proper motion of 5.2 seconds of arc per year. This was, at the time, the largest proper motion known.[2]

Bessel judged that 61 Cygni, despite its lack of brilliance, must be close. He measured the distance between 61 Cygni and each of two very dim (and therefore, he hoped, very distant) neighbor stars with a new instrument called a heliometer which was capable of making very accurate measurements of angular distance. He continued this for over a year.

Eventually, all three astronomers succeeded in determining the parallax of the stars they were studying. Bessel was the first to announce his results, in 1838. Henderson, who had completed his work even earlier than Bessel, waited for his return to England and did not announce his results until 1839. Struve joined in 1840.

It turned out, following certain improvements in the original measurements in later years, that Alpha Centauri (really a triple star, with two members of the system sizable and very close to each other, and a third very distant and very dim companion) has a parallax of 0.760 second of arc. This parallax, only three-fourths of a second of arc, has turned out to be the largest one ever found. Therefore the Alpha Centauri system is the closest known system to ourselves outside our Solar system. It is 4.29 light-years away.

As for 61 Cygni, that, too, is a double star. Its parallax turned out to be 0.29 second of arc, and its distance 11.1 light-years. Vega was the hardest of the three targets to handle because it was the most distant. It is about 27 light-years away.

With parallaxes determined, a second unit of distance came into vogue, one that was introduced by the English astronomer, Herbert Hall Turner (1861–1930). The distance at which a star would have a parallax of one second would be a convenient unit to use. This unit is a "parallax-second," universally abbreviated to "parsec." One parsec is equal to 3.26 light-years, or 200,000 astronomic units, or 19 trillion miles, or 30 trillion kilometers.

The distance to some of the nearer stars is tabulated as follows:

[2]   It is still the largest proper motion known among the stars visible to the naked eye. The half-dozen or so stars that have been found to move still more rapidly are all so dim as to be visible only in telescopes — but they are dim because of their tiny size, not because of their great distance.

Star	Distance Light-years	Parsecs
Alpha Centauri	4.29	1.32
Barnard's star	5.97	1.84
Wolf 359	7.74	2.38
Sirius	8.7	2.67
61 Cygni	11.1	3.42
Procyon	11.3	3.48
Kapteyn's star	12.7	3.87
Van Maanen's star	13.2	4.06
Altair	15.7	4.82

It is clear then that the Solar system is isolated within a vast emptiness. The rough estimates made earlier in the chapter turn out to have been overconservative. Barnard's star, estimated to be ten trillion miles distant on the assumption that it moves as quickly as the Earth, actually moves more quickly and is thirty-five trillion miles away. Sirius, estimated to be ten trillion miles distant on the assumption that it is as luminous as the Sun, is actually considerably more luminous and is fifty trillion miles away. Even the very nearest star, Alpha Centauri, is 25 trillion miles away.

The nearest star is nearly 7000 times as far from the Sun as Pluto is. Imagine a circle drawn with the Sun at the center and Alpha Centauri at the circumference. If this is scaled down to the point where the radius is 9 yards in length (so that a good-sized house can be placed within it), the orbit of Pluto about the Sun will be a tiny ellipse with an extreme diameter of 1/10 of an inch.

Once we know the actual distance of a star, we can calculate its actual luminosity from its apparent magnitude; or, conversely, determine how bright it would appear at any distance. The brightness of a star at the arbitrary distance of 10 parsecs (or 32.6 light-years) is called its "absolute magnitude."

If the Sun were at a distance of 10 parsecs (instead of at 0.000005 parsec, as it is), it would have a magnitude of 4.9 and would be only a dim star. If Sirius were placed at 10 parsecs instead of the 2.67 parsecs it actually is, it too would appear dimmer than it does, but not by very much. Its absolute magnitude is 1.4.

At equal distances from us, Sirius would be 3.5 magnitudes brighter than the Sun. Since 1 magnitude means a ratio of 2.512 in brightness, we can say that Sirius is $(2.512)^{3.5}$ or twenty-five times as luminous as the Sun.

Figures on the absolute magnitude and luminosity of some well-known stars are given below:

Star	Absolute Magnitude	Luminosity (Sun = 1)
Sun	4.9	1.00
Procyon	2.7	7.6
Altair	2.3	10.9
Sirius	1.4	25.0
Vega	0.5	57.5
Arcturus	— 0.3	120
Capella	— 0.3	120
Regulus	— 0.7	173
Aldebaran	— 0.8	190
Canopus	— 3.1	5,200
Beta Centauri	— 5.2	12,000
Antares	— 5.4	13,000
Rigel	— 7.1	25,000
Deneb	— 7.1	25,000

In other words, the Sun, which is the most glorious object in our heavens and which Copernicus thought to be the center of the Universe, is not only merely a star but merely an ordinary star. There are other stars that are thousands of times as luminous as the Sun.

However, we need not be too abashed. While the Sun is not the brightest star, neither is it the dimmest.

Indeed, of the fifty stars closest to the Sun, only three — Sirius, Procyon, and Altair — are distinctly more luminous than the Sun. The two larger stars of the Alpha Centauri system (Alpha Centauri A and Alpha Centauri B) are each about as luminous as the Sun. The remaining forty-five are all dimmer than the Sun and some are much dimmer.

4.

The Galaxy

Olbers' Paradox

By 1840, astronomers had finally plumbed the distance of the stars, at least of the nearest ones, and found them to be something more than a parsec distant.

The next question, a rather inevitable one, is: Where do the stars end? How far is the farthest star? After all, the Earth has a finite surface, the Solar system takes up a finite region of space. In graduating to a new "plateau," are we still in the realm of the finite? Or must we finally find ourselves face to face with the infinite — that concept that has troubled scholars from the beginning.

If we restrict ourselves to that portion of the Universe that can be seen with the unaided eye, the Universe is certainly finite. It is now known that, in the region nearest ourselves at least, the average distance between stars is about 3 parsecs (or 10 light-years). We also know that there are about 6000 stars visible to the naked eye. Suppose that those stars are all that exist and that they are all separated by the average distance. In that case, a sphere with a diameter of about 100 parsecs, or 330 light-years, would enclose all 6000 stars.

This is large enough, certainly, by any ordinary human standards. A sphere 100 parsecs across has a diameter of nearly two quadrillion (2,000,000,000,000,000) miles, and its size would have shocked and flabbergasted any astronomer who lived before 1600 — and many who lived afterward.

However, 6000 stars are not nearly all there are. As soon as Galileo turned his first telescope on the heavens in 1609, he found large numbers of dim stars that were invisible to the unaided eye. Every time the telescope was improved, a new crop of still more copious and still dimmer stars was uncovered.

There seemed, at first, no sign of any end, and in 1800 one might well have veered from a Universe of 6000 stars and a 100-parsec diameter to one of an infinite number of stars and no final boundary at all. In the latter case, the question "How far is the farthest star?" would have no answer because one might say, "There is no farthest star!"

As usual, though, the thought of the infinite was repulsive to mankind. The attack on a possible infinite universe of stars was carried on along two fronts, one of theory and one of observation.

The theoretical reasons for doubting the existence of an infinite universe of stars arose from suggestions made by the German astronomer Heinrich Wilhelm Matthäus Olbers (1758–1840). In 1826, he advanced what has come to be called "Olbers' paradox." To explain that, let us start with the following assumptions:

1. The Universe is infinite in extent.

2. The stars are infinite in number and evenly spread throughout the Universe.

3. The stars are of uniform average luminosity throughout the Universe.

Imagine the Solar system at the center of such a Universe and consider that Universe to be divided into thin, concentric shells like those making up an onion.

The volume of such thin shells would increase as the square of the distance. If Shell A were three times as far from us as Shell B, Shell A would have 3^2, or nine times the volume of Shell B. If the stars are evenly spread through all the shells (Assumption 2 above) then Shell A with nine times the volume would have nine times the stars that Shell B would have.

On the other hand, the light of the individual stars would fall off as the square of the distance. If Shell A is three times as distant as Shell B and contains 9 times as many stars as Shell B, each individual star (assuming uniform average luminosity throughout — Assumption 3 above) in Shell A is only (⅓)² or 1/9 as bright as the individual stars in Shell B.

We conclude then that Shell A has nine times as many stars as Shell B, and that each star in Shell A is 1/9 as bright as each star in Shell B, so that the total light delivered by Shell A to the Solar system is 9 × 1/9 that of Shell B. In short, Shell A and Shell B deliver an equal amount of total light to the Solar system.

The same can be argued for every other shell. It follows then that if there are an infinite number of shells (Assumption 1 above), the total light reaching us would be infinite, except that the nearer stars would block the light of the farther ones. Even allowing for such shielding, the whole sky would glow like the surface of one huge, bright sun. Yet this certainly is not the case.

Olbers suggested that one way out of this paradox might be found in the existence of dust clouds in space, clouds that

Olbers' paradox

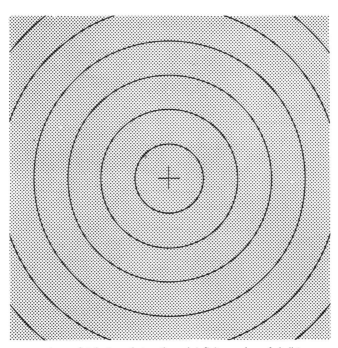

(even distribution of stars through infinite number of shells)

absorbed the light from really distant stars, so that only the light from the relatively nearby stars reached us. This, however, is not good enough. If the dust clouds absorbed the light, they would gradually heat up until they were emitting as much light as they were absorbing. The amount of light reaching us would still be infinite.

There must be something wrong with Olbers' assumptions. The Universe must not be infinite in extent, or, if it is, the stars must not be infinite in number. One would expect, instead, that there would be a finite (although very large) number of stars, spread out over a finite (but very vast) space.

This conclusion, based on Olbers' reasoning, jibed perfectly well with the careful astronomic observations being carried on at this time by William Herschel.

Herschel's Lens

At least one of the assumptions behind Olbers' paradox seems shaky on the face of it.

We might assume that the stars are evenly spread throughout space, but certainly observations of that portion of the Universe that can be seen from Earth do not seem to support this view.

Stretching completely around the sky through the constellations Orion, Perseus, Cassiopeia, Cygnus, Aquila, Sagittarius, Centaurus, and Carina is a softly luminous band that cuts Earth's equatorial plane at an angle of 62°. It fades out in the bright lights of a modern city, but on a moonless night deep in the country, it is a beautiful sight.

The ancients, who did not possess the dubious blessing (from an astronomer's point of view) of the electric light, were well aware of this luminous band. To them it looked like a milky cloud. The Greeks called it "galaxias kyklos" ("milky circle") and the Romans called it "via lactea," which translates directly into the English "Milky Way." From the Greek version of the name comes the English "Galaxy."

In 1610, Galileo looked at the Milky Way through his primitive telescope and found it to be not a featureless luminous cloud but a vast collection of very dim stars, as, indeed, some philosophers of pretelescopic days had speculated it might be.

It seems clear then that there must be many more stars in the direction of the Milky Way than in any other direction. Indeed, even the bright visible stars occur in somewhat greater profusion in the direction of the Milky Way than in other directions. This certainly goes against the assumption of evenly spread-out stars.

Herschel was well aware in his systematic survey of the heavens that there were more stars in some directions than in others, but he was not satisfied with a merely qualitative judgement. In 1784, he decided to count the stars and see exactly how they varied in profusion from place to place.

To count all the stars all over the sky was, of course, an impractical undertaking, but Herschel realized it would be quite proper to be satisfied with sampling the sky. He chose 683 regions, well-scattered over the sky, and counted the stars visible in his telescope in each one. He found that the number of stars per unit area of sky rose steadily as one approached the Milky Way, was maximal in the plane of the Milky Way and minimal in the direction at right angles to that plane.

How could this be explained? Possibly the stars just happened to be spaced more closely as one approached the Milky Way. But why should this be so? There seems no simple way of explaining why this steady change in spacing should occur. It seemed to Herschel much more reasonable to suppose that the stars were spread out with uniform separations, but over a volume of space that was not spherical and therefore not symmetrical in all directions.

Suppose the stars were evenly distributed through a volume of space shaped like a lens or a grindstone and that our own Sun were near the center of the mass. If we sighted along the long diameter of the grindstone, we would see a number of bright stars close to us, and behind them large masses of distant and therefore dim stars, and still farther beyond even more numerous masses of still more distant and still dimmer stars and so on. The numerous stars we would see in the far distance would be too dim to be visible individually, but *en masse* they would lend a pale milky luminosity to the sky and make up the Milky Way.

On the other hand, if you looked away from the long diameter of the grindstone, you would look through smaller and smaller thicknesses of stars. You would then see only the nearby rather bright stars but beyond them would be no mass of distant stars and no milky luminosity.

To Herschel, then, it seemed that the stars of the Universe formed a finite "sidereal system" possessing a definite shape. (Gradually, the word Galaxy came to mean the entire sidereal system, rather than merely the visible Milky Way, so that we could speak of our Sun as one of the stars of the Galaxy.)

From the number of stars he could see in the various directions and making use of his assumption as to even separation, Herschel even tried to make some rough decisions as to the size of this sidereal system. He suggested that the long diameter of the lens-shaped Galaxy was about 800 times the average distance between two stars (which he took to be the distance between the Sun and Sirius—and here he happened to be correct, at least for those stars in the neighborhood of the Sun). The short diameter of the Galaxy he took to be 150 times the mean distance.

What Herschel was envisaging, then, was a Galaxy that could hold 300,000,000 stars, 50,000 times as many as could be seen with the naked eye. Moreover, if the average distance between stars is taken as 10 light-years (although it must be remembered that actual stellar distances were not determined till sixteen years after Herschel's death), the Galaxy was 8,000 light-years in its long diameter and 1,500 light-years in its short.

Furthermore, since the Milky Way seemed to encircle the sky and to be at least fairly bright on all sides, it seemed reasonable to suppose that the Sun was somewhere near the center of the Galaxy.

Herschel's observations and Olbers' reasoning effectively killed the notion of an infinite Universe for a century. The labors of the nineteenth century astronomers to count and record the stars in greater numbers and with greater accuracy merely succeeded in refining the details of Herschel's overall picture.

Lens-shaped star system

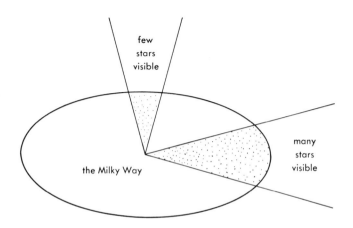

few stars visible

many stars visible

the Milky Way

Laborious star-counting by eye reached its climax with the star-map called the *Bonner Durchmusterung* ("Bonn Survey"), which began to be published in 1859 under the supervision of the German astronomer Friedrich Wilhelm August Argelander (1799–1875) who worked at the University of Bonn. This map eventually recorded the position of about half a million stars.

However, by the second half of the nineteenth century the necessity of pinpointing each star by eye was passing. Photography was developed and applied to astronomy. A photograph of a star-field froze that section of the sky in perpetuity and allowed star counts to be made in comfort and at one's convenience.

One who made great use of photographed star-fields was the Dutch astronomer Jacobus Cornelius Kapteyn (1851–1922). Like Herschel, he sampled the sky. He went further in one respect. He attempted a systematic count of the stars for each increasing magnitude.

If the number of stars was infinite, there should be a steady increase in the total number present in each succeeding shell of space about us (if we revert to the picture we drew in connection with Olbers' paradox, see page 44), since each successive shell as we move outward is larger than the one within and can hold more stars. Since stars, generally appear dimmer with distance, there should be a steady increase in the number of stars with decreasing brightness.

Kapteyn observed, however, that the rate of increase was not steady but began to fall off in the higher magnitudes. This meant that the stars were beginning to thin out in the very distant shells, and Kapteyn was able to make a rough estimate as to the distance of those final shells in which the stars finally petered out.

His results confirmed Herschel's view of a lens-shaped Galaxy with the Sun at or near the center. Kapteyn's figures for the dimensions of the Galaxy were higher than Herschel's, however. In 1906, he estimated the long diameter of the Galaxy to be 23,000 light-years and the short diameter to be 6,000 light-years. By 1920, he had further raised the dimensions to 55,000 light-years and 11,000 light-years. This final set of dimensions involved a Galaxy with 475 times the volume of Herschel's.

The Moving Sun

The sidereal system pictured by Herschel represented another blow at man's estimate of his own importance.

In ancient times, man tended to accept himself very literally as the hub of the Universe. The Universe was not only geocentric, with the Earth — man's home — the immovable center of everything; it was homocentric, with man the measure of all things.

Once Copernicus had done his work, and the heliocentric theory was put forth and slowly accepted, it became more difficult to attach the proper importance to man. After all, he inhabited only one planet of many, and man's planet, moreover, was not the largest nor the most spectacular by far. Earth could not begin to compare with Jupiter for size or with Saturn for beauty.

Nonetheless, to the astronomers of the seventeenth and eighteenth century, the Sun seemed the immovable center of the Universe, and the Sun still belonged to us. It was the source of light and heat, the fount of life on Earth.

And yet, as the notion of the solid vault of the sky faded out, it came to seem less and less likely that the Sun could have this overwhelming importance. With stars spread throughout a vast region and the Sun itself only a star among stars, why should the Sun be the center of the Universe any more than the Earth was?

Furthermore, if one accepted a Galaxy of the size suggested by Herschel, one that was populated by hundreds of millions of stars, how could one seriously maintain that our Sun counted for very much among so many spread out so far.

As the proper motions of more and more stars were observed, there seemed no indication that the stars generally were moving in some grand revolution about the Sun. The motions seemed to be random in nature, so it became more and more tempting to suppose that all stars were moving in this more or less random fashion (like individual bees in a large swarm) and that those stars which showed no proper motion were either too far away to show one until they had been observed for many more centuries, or else happened to be moving directly toward us or away from us, so that a crosswise proper motion could not be detected.

If this were so, then it was reasonable to suppose that the Sun was moving, too. Why should it alone be motionless in a Universe of moving stars? Herschel reasoned thus in 1783 and set about trying to determine what the Sun's motion might be.

Suppose the Sun were surrounded by stars spread evenly through space. Those near the Sun would be seen as separated by comparatively large distances, while those far away would seem to be much closer together. We see the same phenomenon if we look at the regularly spaced trees of an orchard, or men standing in carefully ordered ranks and files. It is a common effect of perspective.

If a group of stars, then, were brought closer to the Sun, without any alteration of their positions relative to each other, they would seem to have become more widely spread apart. Therefore, in the process of approaching, they would seem to be moving apart. On the other hand, if a group of stars receded, they would, by the same token, seem to be moving together.

If the Sun were moving through the Galaxy, the stars lying ahead of it in its direction of motion would seem, on the whole, to be approaching the Sun and therefore to be moving apart. (This effect would be masked in part by the fact that the stars are not standing still but have motions of their own in all directions—but it would not be completely masked.) The stars lying behind the Sun in the direction opposite that in which it is moving would seem, on the whole, to be receding from the Sun and therefore to be moving together. Finally, the stars lying at right angles to the direction of the Sun's motion would have the largest proper motions and these would tend, on the whole to be in the direction opposite that taken by the Sun.

Motion of the Sun

Apex

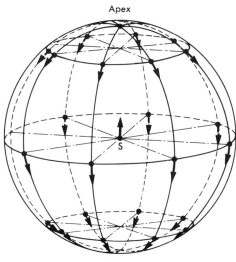

Antapex

Not many proper motions had been determined in Herschel's time, but by making use of those that were known, Herschel was able to suggest on the basis of this reasoning that the Sun was indeed moving. It was moving toward a spot in the constellation Hercules.

This proved to be not too bad an estimate. Many more proper motions have been determined in the century and a half following Herschel's work, but the "Apex" (the point toward which the Sun seems to be moving) is now considered to be not far from the point Herschel determined. According to the best observations now available, it seems to be located in the constellation of Lyra, which is next to Hercules. The Sun is moving toward the Apex (relative to the closer stars) at a velocity of twelve miles per second.

Star Clusters

The fact, then, that the Sun seems to be at or near the center of the Herschel-Kapteyn model of the Galaxy is not to be considered of great significance. It would seem to be merely a fortuitous circumstance; had mankind made its astronomical observations during another age, far in the past or far in the future, it might have found itself located toward one end or the other of the Galaxy.

It is not much of a sop to human vanity to think that the Sun (and therefore the Earth and man himself) is at the center of things only as a matter of accident. Yet even this much grew shaky and uncertain as Kapteyn polished the final details of his model in the first two decades of the twentieth century.

The Herschel-Kapteyn model was in trouble as the result of evidence that arose in connection not so much with stars individually as with groups of stars.

Even to the naked eye, such groups seem to exist. The best known group is the "Pleiades," a small cluster of moderately bright stars in the constellation of Taurus, the Bull. There are nine stars in the cluster that are bright enough to be made out by the unaided eye, but some are too closely spaced to be made out separately. The average eye can see six or seven. (The cluster is sometimes called the "Seven Sisters.")

When Galileo turned his telescope on the Pleiades in 1610, he found he could easily count 36 stars in the group, and modern photography shows at least 250, with the total count probably close to 750.

The Pleiades represent a true association of stars and **not** merely an accidental view of a number of stars at varying distances in nearly the same line of sight. Bessel demonstrated this in 1840 when he showed that the various members of the cluster had proper motions of 5.5 seconds of arc per century in the same direction. If they were independent stars, it would be entirely too much to ask of coincidence that they all be moving in the same direction and at the same velocity.

Astronomers have estimated that the average distance between stars in the Pleiades cluster is only one-third that of the average distance between stars in our own neighborhood. The whole group is now known to be some 400 light-years from us and to be spread out over a region of space some 70 light-years in diameter.

Although the Pleiades are the most beautiful cluster visible to the naked eye, they are only the feeblest token of the spectacles made visible by the telescope.

Without knowing it, the French astronomer Charles Messier (1730–1817) had a glimpse of these greater glories while searching for considerably lesser objects. Messier was a comet-hunter and, in his lifetime, discovered quite a few. However, he grew tired of reacting to fuzzy objects in the sky which were permanently fixed and were *not* comets. In 1781, he made a careful map of about forty such objects in order that he, and other comet-hunters, might know their location and learn to ignore them. Eventually, Messier and others increased the number of objects on the list to a hundred.

Among them, for instance, was one fuzzy starlike object that had first been observed by Halley in 1714. Because it was thirteenth on Messier's list, it is sometimes called M13. When William Herschel studied M13 several decades later, with a much better telescope than the one that had been available to Messier, he realized he was not looking at a mere blur of light but at a densely packed spherical conglomeration of stars.

The Pleiades consisted of a group of comparatively widely spaced stars and was therefore called an "open cluster," whereas the object, M13, was a closely spaced globe of stars and was therefore called a "globular cluster." Because M13 is in the constellation of Hercules, it is now frequently called the "Great Hercules Cluster." A globular cluster consists not of hundreds of stars but of thousands. Some 30,000 stars have actually been counted in the Great Hercules Cluster, and the total must be well over 100,000, perhaps even close to a million. Toward the center of the cluster, stars must be distributed with an average separation of considerably less than 1 light-year.

Nor is this the only globular cluster. Several others are included on Messier's list, including M3 in the constellation Canes Venatici (the Hunting Dogs) and M22 in the constellation Sagittarius (the Archer). About a hundred such globular clusters are known and it is estimated that three hundred may exist in our Galaxy altogether.

Oddly enough, the globular clusters are not distributed evenly over the entire sky—something first remarked on in the early nineteenth century by John Herschel (1792–1871), son of the famous William Herschel, and himself an astronomer of note.

Almost all the globular clusters are located in one hemisphere of the sky and, indeed, a third of them are crowded into the single constellation of Sagittarius, which makes up only 2 percent of the sky. John Herschel believed this could not be accidental, but had to have some significance.

What that significance might be eluded astronomers for the next century, partly because the actual position in space of these globular clusters was not known. They were far too distant to have a measurable parallax, and until the twentieth century, no other method for determining stellar distances was known.

Variable Stars

A proper method of handling the problem of globular clusters was found in the early twentieth century in connection with a certain variety of "variable star." (A variable star is one that periodically varies in brightness.)

The history of variable stars is surprisingly short, considering that a few stars bright enough to be seen by the unaided eye vary appreciably in brightness. No comments concerning such variations in brightness have come down to us from the ancient astronomers. Indeed, the Greek view, as expressed by Aristotle, was that everything in the heavens was permanent and unchangeable. This weighty official view did not, however, alter the fact that variable stars exist and that they attest to the fact that there *is* change in the heavens.

The most notable variable among the naked-eye stars is Beta Persei, second brightest star in the constellation of Perseus. It dims and brightens quite perceptibly. Every two days and twenty-one hours, it rapidly loses a little more than a full magnitude of brightness for a brief period, and then as rapidly regains the loss.

Neither the Greeks nor the Arabs (the latter being the great astronomers of the early Middle Ages) made mention of this although individuals may have noted it and been uneasy about it. The Greeks, in their fanciful description of the stars of this constellation, pictured Beta Persei as located in the head of Medusa, which Perseus is holding. Medusa was a demon with living, hissing snakes for hair and a face so horrible that anyone who beheld it turned to stone. Beta Persei was sometimes called the "Demon Star" because of this, a sign that something might have been considered to be wrong with it. Then, too, the Arabic name for the star is Algol ("the ghoul"), which again indicates something chilling.

In 1782, the English astronomer John Goodricke (1764–1786) studied Algol in detail and suggested that a dim star was circling it with its orbit nearly edge-on to the Earth. Every time it circled Algol, it passed between that bright star and the Earth and cut off part of its light. The suggestion was not taken seriously at the time, but additional evidence since then has shown it to be almost certainly correct. Algol is now accepted as the best-known example of an "eclipsing binary." Its light does not really wax and wane; it is merely regularly blocked off.

Indeed, the system is even more complicated than it would seem. The two stars that eclipse each other are two million miles

Light-curves of eclipsing binaries

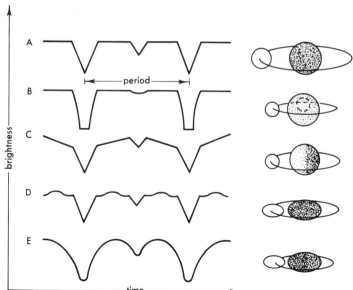

apart, but a third star at a considerably greater distance revolves about the two with a period of 22 months.

Light variability due to eclipses has nothing to do with the internal structure or properties of a star. If the companion of Algol revolved about Algol in another plane and did not intercept its light on the way to the Earth, Algol would not be considered a variable.

Quite different is the case of the star Omicron Ceti, located in the constellation of Cetus (the Whale). This star was first carefully observed in 1596 by the German astronomer David Fabricius (1564–1617). At its brightest, this star may attain a magnitude of 2 and at its dimmest it becomes too dim to be seen by the naked eye. It received the name of "Mira" ("wonderful") as a result. Its period of variation is about eleven months—that is, its period of peak brightnesses come about eleven months apart. This is rather long for a variable star, and Mira is therefore said to belong to the class of "long-period variables."

Mira, unlike Algol, is truly variable. It actually grows dimmer and brighter, astronomers have concluded. It is therefore also classified as an "intrinsic variable."

Another example of an intrinsic variable is Delta Cephei, the fourth brightest star in the constellation Cepheus. It differs from Mira considerably. In the first place, Delta Cephei's period of variation is short, 5.37 days, and in the second, it is regular.

Other variable stars like Delta Cephei have been discovered, each with short and regular periods of variation. The periods range from two to forty-five days, with periods in the neighborhood of a week being very common. The manner in which the brightness increases and decreases is distinctive, too, and all these stars are grouped together as "Cepheid variables" or "Cepheids," after the name of the first to be studied.

Although the Cepheids were interesting stellar curiosities, they did not seem at first to have any great significance. That view changed sharply in 1912, when the American astronomer Henrietta Swan Leavitt (1868–1921) began to locate and systematically study hundreds of Cepheid variables in the Small Magellanic Cloud from an observatory set up by Harvard at Arequipa, Peru.

The Small Magellanic Cloud is one of two areas of luminosity (the other is the Large Magellanic Cloud), which look like isolated patches of the Milky Way. They are located so far south as to be invisible to observers in the North Temperate Zone. They were first described in 1521 by the chronicler accompanying Magellan's voyage of circumnavigation of the globe—whence their names.

The Magellanic Clouds were not studied in detail till 1834 when John Herschel observed them from the astronomic observatory at the Cape of Good Hope. Like the Milky Way, they represented assemblages of large numbers of dim stars—dim, presumably, because of great distance.

In fact, the Clouds are so far from us that variation in distance between stars on the near side and on the far side can be expected to be relatively unimportant. As an analogy, we can say that all the people in Chicago are about the same distance from Times Square in New York; the people in eastern Chicago are a little closer than those in western Chicago, but the difference is too small to be significant in comparison with the total distance.

This means that all the stars and, in particular, all the Cepheids in the Small Magellanic Cloud may be taken as being about the same distance from us. If one Cepheid in the Cloud seems to be brighter than another, it can only be because it is actually brighter—more luminous—than the other. No artificial difference in brightness need be expected to be imposed on them because one is much closer than another.

In her studies of the Cepheids of the Small Magellanic Cloud, Miss Leavitt noted that the brighter the Cepheid variable, the longer its period. A Cepheid in the Cloud with a magnitude of 15.5 had a period of two days; one with a magnitude of 14.8 had a period of five days; one with a magnitude of 12.0 had a period of 100 days. Apparently, there was some regular relationship between luminosity and period.

This relationship ought to be true of the Cepheids of our own neighborhood, too, as well as for those in the Magellanic Clouds. (Scientists generally find it profitable to assume that a relationship that holds in one place or under one set of conditions also holds in another place or under another set of conditions—at least until there is evidence to the contrary.) Why, then, was it not evident?

The trouble is that the period is related to the luminosity of the Cepheid, and in our own neighborhood the luminosity may be masked as a result of distance. A very luminous Cepheid with a long period may be so far away as to appear faint, whereas a much less luminous Cepheid with a short period may be quite close to us and appear bright. In such a case, bright stars can appear to have short periods and dim stars long ones. Indeed, the confusion of distances makes it seem that there is no connection whatever between brightness and period. As a matter of fact, there is not. The connection is between *luminosity* and period, and from the apparent brightness of a star, one cannot tell the actual luminosity unless one also knows the distance of a star.

Unfortunately, in 1912 the distance of not one single Cepheid was known. Only the measurement of parallax could be used to determine distances and that had its limitations. The farther the star, the smaller the parallax, and the more difficult it is to measure. Even today, the parallactic method is impractical for distances greater than 150 light-years, and there is no Cepheid variable as close to us as that. The closest is at least 300 light-years away.

Miss Leavitt noted the luminosity-period relationship in the Small Magellanic Cloud not because she knew the distance of the Cloud but because within the Cloud distances did not matter. Brightness within the Cloud was proportional to luminosity, and therefore the luminosity-period relationship showed up as an easily observed brightness-period relationship.

Once this relationship had been discovered, however, it could be applied to our neighborhood and used as a yardstick for distances far beyond those that could be plumbed by parallax.

Suppose, for instance, that two Cepheids were observed to have equal periods but that one seemed brighter than the other. Their luminosity would have to be the same since their periods were the same, so the difference in apparent brightness would be entirely the effect of distance. It would be easy to calculate how much more distant one Cepheid would have to be than the other to account for the difference in brightness.

If two Cepheids had different periods, the difference in luminosity could be calculated. The difference in apparent magnitude could be measured directly, and from both pieces of information the relative distances could be calculated.

Such determinations of relative distances merely gave a plot of the Galaxy to scale (as Kepler's laws had once done for the Solar system) and did not give actual distances. Even this much had its value, however.

In the years after Miss Leavitt's discovery, the American astronomer Harlow Shapley (1885–1972) made use of the Cepheid scale to study the globular clusters. Many clusters contained some Cepheid variables. By measuring the periods of these, Shapley could determine their relative luminosity. By comparing this with their apparent brightness, Shapley could determine their relative distances—and the relative distances of the globular clusters of which they formed a part. When clusters contained no clearly seen Cepheids, he measured the apparent magnitude of the brightest stars of the entire cluster, assuming these to be roughly the same in intrinsic brightness in all cases.

When he did this, he found that the globular clusters seemed to be distributed in a spherical arrangement. They marked out a large ball with its center in the direction of the constellation Sagittarius. Astronomers on Earth viewed this ball of globular clusters from the outside—and from far outside, too, so that the entire structure seemed to occupy a relatively small portion of the sky in and about Sagittarius.

Why did the globular clusters arrange themselves so? To Shapley it seemed logical to suppose that the clusters were grouped about the massive center of the Galaxy, as the planets are grouped about the Sun, the massive center of the Solar system. If so, the Galactic center is far from the Sun, and we are at the outskirts of the Galaxy rather than at its center.

If we only knew how far the sphere of globular clusters actually was from us, we could then determine how far the Galactic center might be, how far out in the outskirts we were, and so on. If the distance of even a single Cepheid were actually known, the distance of all the rest could be calculated. The actual dimensions of the Galaxy (and possibly of the Universe, if the Galaxy were all there were to the Universe) could be solved.

But how can the distance of the Cepheids be determined when not one is close enough to have its parallax measured?

To explain Shapley's method for circumventing this purely fortuitous bad break, I will have to engage in a rather long digression and go back in time some three-quarters of a century.

Distribution of globular clusters

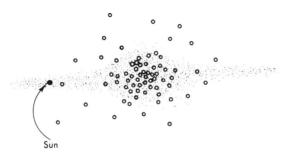

Sun

5

The Size of the Galaxy

The Doppler Effect

Those of us who lived in the days when trains were more common than they are now know that the whistle of an approaching train is higher pitched than the whistle of a train standing still, relative to us.[1] Similarly, if a train is receding from us, the pitch of the whistle is lower than it would be if it were standing still, relative to us. If we were waiting at a station and a train, sounding its whistle, approached us, passed us without stopping, and then hastened away, the pitch of the whistle would drop suddenly as it passed.

In 1842, this phenomenon was explained accurately by an Austrian physicist, Christian Johann Doppler (1803–1853).

To begin with, sound is a series of compressions and rarefactions of air. The distance from one region of compression to the next is equal to the wavelength of the sound. The longer the

[1] A train does not have to be absolutely motionless to be standing still, relative to us. If the train were moving and we were on it, moving with it, the train would be standing still, relative to us. As long as it is not speeding up, slowing down, or negotiating a curve, but is moving at a constant velocity, however high, we can still walk about in it as if it were standing still. And it would indeed be standing still, relative to us, even if it were not standing still, relative to the surrounding countryside.

wavelength, the deeper the pitch of the sound we hear; the shorter the wavelength, the higher the pitch.

Suppose a train whistle, sounding at a constant pitch, is stationary with respect to you. A region of compression is produced and spreads outward, followed by another such region, then by still another, and so on. The regions of compression are separated by some fixed distance.

If, however, the train and its whistle were moving toward you, the second region of compression is emitted a little closer to you than was the first. The train has moved toward the speeding first region of compression, so that the second region is closer to the first than it would have been if the train were standing still. The same thing happens in the case of the third region and the fourth. As long as the train keeps approaching you, it gains slightly on the sound waves and the regions of compression are consistently closer together than they would be if the train were standing still. The wavelength of the sound is therefore shorter, and a higher pitch is produced than would be if the train were standing still.

Precisely the reverse takes place when the train is receding from you. The second region of compression is produced farther from you and from the first region; the third region is produced still farther away, and so on. The wavelength of the sound produced by a receding whistle is therefore long and its pitch deeper than it would be if the train were motionless.

The faster a train speeds toward you, the more closely spaced are the sound waves and the higher the pitch; the faster it speeds away from you, the more widely spaced are the sound waves and the lower the pitch. Knowing the normal pitch of the whistle and the pitch one hears, it would be possible to determine whether the train were approaching or receding and at what velocity, without requiring any other evidence.

This change of pitch with motion is called the "Doppler effect" in honor of the physicist.

The Spectrum

In theory, the same effect should be noted in the case of any wave form radiating outward from a source. Notably, it would be detected in the case of light, as Doppler himself pointed out.

Light, like sound, is a wave form (although not of the same type). Light, too, possesses wavelengths, and differences in these wavelengths can also be detected by the senses—as a difference in color. The longest visible wavelengths are seen as red. As wavelengths grow shorter, the colors change to orange, yellow, green, blue, and violet, in that order. The colors are not differentiated sharply, of course, but gradually shift from one into the other. We see this effect in nature in the rainbow.

Isaac Newton was the first to study a man-made rainbow in detail. In 1666, Newton allowed a beam of sunlight to enter a darkened room through a hole in a window-blind and then pass through a triangular piece of glass or "prism." The light beam was bent, or "refracted," on passing through the prism and fell on the wall opposite as a broadened spot of successive colors, similar in appearance to the rainbow. Newton called the band of colors a "spectrum."

In this way, Newton showed that sunlight was not a pure unmixed entity, for if it had been it would all have been refracted through the prism in the same way and would have struck the wall as an unchanged beam of white light. Instead, it seemed actually to be a mixture of a large variety of kinds of light, each type being refracted through a slightly different angle and each being interpreted by our sense of vision as a different color. The mixture of all the different varieties, in the proportions that occur in sunlight, is sensed by us as white light.

We now recognize that the varieties of light in sunlight are distinguished from each other by wavelength. The extent to which a ray of light is refracted by glass depends on its wavelength; the shorter the wavelength, the greater the refraction. Orange light is refracted more than red light, yellow light is refracted to a still greater extent, and so on. Violet light is refracted most of all. In the final spectrum, the red light is at the end that is least refracted, the end that is closest to the direction of travel of the original unrefracted light beam. Violet light is at the other end.

Within any one of the colors—orange, for example—the longest, most nearly red, wavelengths are toward the red end of the spectrum, while the shortest, most nearly yellow, wavelengths are toward the violet end of the spectrum.

Suppose, now, that we try to apply the Doppler effect to light.

The Sun, on the whole, is neither approaching the Earth nor receding from it. Therefore, it produces what seems to be a balanced spectrum that contains the mixture our sense of sight interprets as white. However, if the Sun were approaching us, might we not suppose that the wavelengths of light would be squeezed together and that every wavelength of light reaching us would be shorter than it normally would be. There would be a shift of the entire spectrum toward the short-wavelength end. Every bit of the red band would shift toward the orange, every bit of the orange would shift correspondingly toward the yellow, and so on. Because the entire shift is toward the violet end of the spectrum, it is referred to as a "violet shift."

Under such conditions, one might expect the mixture of light in a spectrum no longer to produce a clear white. There would be a deficiency at the red end, an excess at the violet end, and the color of the Sun (if it were approaching us) might be expected to take on a bluish tinge. The more rapidly it approached us, the bluer its light would become by this reasoning.

The same argument, in reverse, could be used to forecast what would happen if the Sun were receding from us. This time the crests of successive light waves would be pulled apart. Wavelengths would become longer than normal, and the entire spectrum would be shifted toward the red end of the spectrum—the "red shift." If the Sun were receding from us, we might reason, its light would take on an orange tinge. The more rapidly it receded from us, the more deeply orange its light.

And yet, although this reasoning seems airtight, it is betrayed by the facts. The trouble is that the light we see in a spectrum is not all there is to the spectrum.

In 1800, William Herschel studied the spectrum produced by sunlight (the "Solar spectrum"). He noted the heating effect produced on a thermometer exposed to different portions of the spectrum, measuring in this way the total energy content of each portion. It would have been natural to expect the temperature rise to become less marked and then vanish completely as one approached the end of the spectrum. This was not so as far as the red end was concerned. Indeed, the temperature rise was greater some distance beyond the red end of the spectrum than anywhere within it.

Herschel suggested that sunlight included wavelengths of light that were longer than any that could be sensed by our eye. Such wavelengths would be refracted still less than those of red light and would be located beyond the red end of the spectrum. This would be "infrared radiation" ("below the red").

Such radiation would be real and would differ from ordinary light only in the size of its wavelength and in the fact that the human eye could not sense it. Thus, by 1800, one could no longer speak merely of "light" to define the entity by which we see; one had to speak of "visible light" instead. Invisible light was no longer a contradiction in terms, but an actual fact (although today infrared light is easily detected by appropriate instruments even though it remains invisible to the eye).

Nor was the violet end of the spectrum a true end. In 1801, the German physicist Johann Wilhelm Ritter (1776–1810) studied the ability of light to bring about certain chemical reactions. Light could, for instance, bring about a breakdown of the white chemical, silver chloride, liberating tiny particles of metallic silver. Small particles of metal usually appear black, and the silver chloride was therefore increasingly blackened as it was exposed to light. Ritter found that different portions of the Solar spectrum were not equally effective in bringing about this change. The shorter the wavelength the more quickly the silver chloride was blackened. Inspired by Herschel's discovery the year before, Ritter tested the region beyond the violet end of the spectrum, where no light at all could be seen. Sure enough, the silver chloride was darkened more rapidly there than it was anywhere within the visible spectrum.

The conclusion was that there were wavelengths of sunlight shorter than any that could be sensed by the eye. Such wavelengths would be refracted still more than those of violet light and would be located beyond the violet end of the spectrum. This would be "ultraviolet radiation" ("beyond the violet").

We must therefore visualize the spectra produced by the Sun (and by other stars) as consisting not only of the red-to-violet colors we can see, but of invisible regions of light beyond the red and violet. If a star is approaching us, so that the wavelengths of its light grow shorter and its spectrum undergoes a violet shift, then light does not "pile up" at the violet end. It "spills out" the end

Beyond the visible spectrum

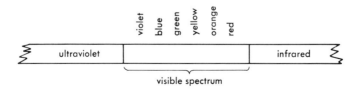

visible spectrum

instead, moving into the ultraviolet region and invisibility. Nor does the shift leave a vacant region at the red end. Infrared radiation turns visible red with the wavelength-shortening and moves in to fill the gap.

The spectrum as a whole shifts, but the visible portion remains unchanged; what it loses at one end it gains at the other. The same argument holds if the star were receding from us to produce a general red shift.

To be sure, a star may be approaching us so rapidly that the entire infrared region is shifted far into the visible or receding from us so rapidly that the entire ultraviolet region is shifted far into the visible. In that way, there would indeed be a visible shift in color toward the violet or toward the red. However, the velocities of approach or recession required to produce such overall shifts in color are so high that it seemed inconceivable (in the nineteenth century, at least) that such speeds would be encountered among the various heavenly objects.

And yet, as it happens, there are indeed differences in colors among the stars. Some stars, Antares for instance, are distinctly reddish in color; others, like Vega, are distinctly bluish. Can this possibly signify that Antares is receding very rapidly from us or that Vega is very rapidly approaching us.

Unfortunately not. There are other causes that might account for differences in color.

Thus, the Sun reddens at the time of sunset although it is not then receding from us. What is happening then is that its light is passing through an unusually great thickness of atmosphere, and the molecules and dust particles in the atmosphere have a greater opportunity to reflect and scatter the sunlight. The short wavelengths of light are more efficiently scattered than are the long wavelengths. Thus, the daylight sky is blue with scattered sunlight from the short-wavelength end of the spectrum. The amount scattered in this fashion when the Sun is high is not enough to affect the color of the sunlight significantly. However, as the Sun approaches the horizon so much of this short-wavelength region is scattered and removed from sunlight by the greater thickness of dusty atmosphere, which the light must pass through, that what is left is distinctly red in color.[2]

Again, a glowing substance will change its color if its temperature is altered. An iron ball, if gradually heated, will even-

[2]   The setting Sun is redder than the rising one, because at the end of the day, the air is generally dustier than it is at the beginning.

tually become hot enough to glow a deep red in color. If the temperature continues to rise, it will glow a brighter red, then orange, then whitish, then blue-white. The higher the temperature, the greater the overall shift of its light in the direction of the shorter wavelength; however, such a shift would not imply any motion of the ball toward us or away from us.

Finally, color may vary with chemical composition. If an object shines by reflected light, this is obvious. Different dyes and different pigments will reflect lights of different colors. Even when a substance is itself the source of light and not merely a passive reflector, color can depend on chemistry. If ordinary table salt (a compound of the metal, sodium) is heated until it glows, it will produce light of a distinctly yellow tinge. Compounds of strontium, when heated, will produce a red light, those of barium a green light, those of potassium a violet light, and so on.

Does it follow from this that there is no way of telling from starlight anything about the motion of a star, despite the Doppler effect? Not at all. In fact, the situation is even better than you would expect; not only motion but many other properties of stars can be determined by examining the spectra they produce. A spectrum contains more than a rainbow of colors—much more.

Spectral Lines

In 1814, the German optician Joseph von Fraunhofer (1787–1826) revolutionized the study of spectra. He was a manufacturer of prisms of fine glass and he tested their quality by their ability to form spectra. Following the work of the English chemist William Hyde Wollaston (1766–1828), Fraunhofer passed a beam of sunlight through a fine slit before passing it through the prism. Each wavelength of light was refracted through a characteristic angle and produced an image of the slit in its own particular color at a particular position on the screen on which the light was thrown. The various images overlapped to produce a nearly continuous rainbowlike spectrum.

The spectrum was not completely continuous because some wavelengths were missing in sunlight and these missing wavelengths showed up as dark lines in the spectrum; slit-images that were not there, so to speak. Such dark lines in the Solar spectrum had been observed by Wollaston, but Fraunhofer's excellent prisms made them much clearer and made many more of them visible—hundreds of them. Fraunhofer was the first to study the dark lines in detail and to map their exact position in the spectrum. They are therefore called "Fraunhofer lines" or, more generally, "spectral lines."

The spectral-line pattern of the Solar spectrum is quite distinctive. Other self-luminous objects, such as the various stars, also show dark lines in their spectra, but in patterns that often differ markedly from that in the Solar spectrum. Nevertheless, certain spectral lines, particularly the most prominent, which Fraunhofer had denoted by letters of the alphabet from A to K, appeared like landmarks in the spectra of most stars.

It is possible to measure the wavelength corresponding to any spectral line by measuring the angle of refraction associated with it; that can be done by accurately measuring its position on the screen against a reference scale. If, for any reason, the wavelength corresponding to the spectral line is shortened, the angle of refraction associated with it is made greater and the position of the line moves toward the violet end of the spectrum. If the wavelength is made longer, the line moves toward the red end of the spectrum.

Soon after Doppler had explained the changing pitch of sound with motion, the French physicist Armand Hippolyte Louis Fizeau (1819–1896) pointed out that to detect the effect in light, one should not worry about overall color, but should measure the exact position of the spectral lines and note their shift.

To see why this should be so, imagine a long featureless rod of which you can only see a small portion. If this rod is shifted slightly, we see a somewhat different portion, but it is still featureless; we cannot tell how much the rod has shifted or in which direction, or, indeed, whether it has moved at all. If, however, the part of the rod we saw contained a marking, then we could easily detect any movement of the rod by noting the shift in position of the marking.

It is in this fashion that spectral lines make it possible to detect any Doppler shift in light. They are the marking on the spectrum. And it is because Fizeau pointed this out that we sometimes speak of the "Doppler-Fizeau effect" in connection with light.

The effect is small, however, and hard to determine. Sound travels relatively slowly—about 331 meters per second—and a train can easily move at one-tenth this speed. Sound waves can then be pushed considerably closer together or pulled considerably farther apart. Light, however, travels at 300,000,000 meters per second, or nearly a million times the speed of sound. Most stars move (relative to ourselves) at less than one ten-thousandth of this speed. The wavelengths of light from such stars are only very slightly altered by their motions.

It was not until 1868 that the British astronomer William Huggins (1824–1910) was able to detect a small shift in the spectral lines of the bright star, Sirius, and to show that it was moving away from the Sun.

It is important in this connection to realize that stellar motions must be understood three-dimensionally. A star might be moving in a direction that is exactly at right angles to our sight or exactly in our line of sight.

Either case is very unlikely. It is much more likely that a star moves neither in the line of sight nor at right angles to it, but somewhere in between. When this is the case, the motion can be split into two components, one in the line of sight and one at right angles to the line of sight. This is done by representing the actual motion as the diagonal of a rectangle. The two adjoining sides of the rectangle then represent the components. The lengths of the diagonal and two sides are proportional to the actual motion and the two component motions, respectively.

The component in the line of sight is called the "radial velocity" because that is the motion toward us or away from us along an imaginary radius that connects our eye to the vast sphere of the sky. The component at right angles to the line of sight is the "transverse velocity" (the velocity "across" the line of sight). It is this transverse velocity that shifts the star bodily across the sky and makes itself evident as the proper motion of the star.

The relative size of the two components depends on the angle the star's motion makes to the line of sight. Conversely, if the size of the radial velocity and the transverse velocity are both known,

Component motions

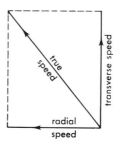

then the actual velocity of the star can easily be deduced.

The two component velocities are determined in two entirely different ways.

The radial velocity is reflected in the star's spectrum by means of the Doppler-Fizeau effect. By measuring the extent of the shift of the spectral lines, the radial velocity (toward us if the shift is toward the violet, away from us if it is toward the red) can be determined directly as kilometers per second. This determination does not depend on the distance of the star. No matter how far the star may be from us, the wavelengths of the spectral lines have set values and a given shift in those values indicates a specific radial velocity whether the star is 4 light-years away or 4000. The only requirement is that the star be bright enough to yield a spectrum in which the position of the lines can be measured.

A star's transverse velocity is not reflected in the spectrum at all, but only as a bodily shift across the vault of the sky. It is therefore detected as proper motion and measured as angles of arc. In order to change a proper motion (in seconds of arc) into transverse velocities (in kilometers per second) one must know the distance of the star. For instance, if Barnard's star moves across the sky at a rate of 10.3 seconds of arc per year and is 6.1 light-years away, we can calculate that it has a transverse velocity of some 90 kilometers per second (or 56 miles per second).

The Galactic Center

Now it is time to return to the problem of determining the distance of the Cepheid variables.

Suppose the matter is considered statistically. In the case of some stars, moving in directions close to the line of sight, the radial velocity is greater than the transverse velocity. In the case of other stars, moving close to right angles to the line of sight, the transverse velocity is greater than the radial velocity. *On the average,* however, these two opposing tendencies tend to cancel, and the radial and transverse velocities may be taken, in general, as equal.

If this is so, then the measured proper motion of a Cepheid (in angles of arc per year) may be taken as equivalent not only to the transverse velocity but to the radial velocity as well, and the radial velocity can be measured directly, by spectral means, as kilometers per second. If you know a motion both as a particular angle of arc per year and as a particular number of kilometers per second, you can calculate the distance; for at one distance, and one distance only, a velocity of so many kilometers per second will produce a shift of a particular angle of arc per year.

For any given Cepheid, a calculation of this sort may give a wildly wrong answer for the distance, for in its particular case, the radial velocity may be much larger or much smaller than the transverse velocity. If a number of Cepheids in a given cluster, all of the same period, are taken, however, and the distances determined for each, then the *average* distance is quite likely to be close to the truth.

The Danish astronomer Ejnar Hertzsprung (1873–1967) determined the distances to certain Cepheids in this manner in 1913, but several years later, Shapley applied this statistical method to the specific problem of determining the structure of the Galaxy. He determined the distances to Cepheids of varying periods. From these distances and the apparent brightness of the Cepheids, he could calculate the luminosities of the Cepheids.

If the luminosities, so determined, were plotted against the periods, then one should get, at least roughly, a smooth line if one allowed for a certain scattering, owing to the intrinsic uncertainty of the method. At least, one·should get this if there were any validity at all to the method. On the other hand, if this statistical method were of no value at all, then the distances determined would be all wrong, the luminosities in consequence would also be all wrong, and there would be no reasonable correlation between luminosity and period.

When Shapley plotted luminosities against period, however, he obtained a very good smooth line, and his results could therefore be accepted as essentially correct. Since he knew both the luminosity and period of representative Cepheids, he could determine the distance of any Cepheid. The Cepheid yardstick had become absolute.

In this way, Shapley could determine the actual distances of the various globular clusters and then go on to calculate the distance of the center of the sphere over which they were distributed. The center of this sphere he assumed to be the center of the Galaxy, and according to his figures it was 50,000 light-years (15,500 parsecs) from the Sun.

By 1920, then, the position of man in the Universe had again been altered, drastically, and once again in the direction of increased humility. Copernicus had shown that the Earth was not the center of the Universe, but he had been certain that the Sun was, as part of the order of nature. Even Herschel and Kapteyn had considered the Sun at the center of the Galaxy (and therefore of the Universe), at least through accident if not through natural order. Now Shapley showed, quite convincingly, that this was not so, that the Sun was far on the outskirts of the Galaxy.

In place of Ptolemy's geocentric Universe and Copernicus' heliocentric Universe, we now had Shapley's "eccentric Universe", one in which the Sun was "away from the center" — the literal meaning of "eccentric."

Shapley's eccentric Universe raised some problems, however. If the bulk of the Galaxy was to one side of the Sun, off in the direction of the constellation Sagittarius, why was not the band of the Milky Way vastly (instead of only moderately) brighter in that direction than in the opposite direction, where only the tag end of the Galaxy existed?

The answer to that question came with the realization that there was more to the Universe than met the eye—quite literally.

Not everything that glitters in the heavens beyond the Solar system is a star. There were also more diffusely bright objects, some of which had been carefully mapped by Messier, and some of which had been observed even before Messier.

In 1694, for instance, the Dutch astronomer Christian Huygens (1629–1695) entered in his diary the description of a bright, fuzzy region in the constellation of Orion. Such a bright, fuzzy region, resembling a luminous cloud, came to be called a "nebula" (the Latin word for "cloud"). the particular one described by Huygens is the "Orion Nebula."

The Orion Nebula is a huge object. It is now known to be about 1600 light-years away, and in order for it to show the diameter we observe it to have, it must be 30 light-years in diameter. It is a vast cloud of dust particles that are very rarefied by earthly standards. It is less than a millionth as dense as the best vacuums we can produce in the laboratory, but it is so voluminous as to surround a number of hot stars, reflecting and scattering their light. Many other luminous nebulae of this sort, some exceedingly beautiful, are now known.

Astronomers became aware, however, that just as there were regions of space that glowed with a soft, cloudy luminosity, so there were other regions in which there was a surprising lack of luminosity. Thus, William Herschel, in studying the Milky Way closely, noted regions where there were very few stars, although they might be bounded by other regions that were simply bursting with vast numbers of stars. Herschel took these dark areas at face value, assuming that they represented regions that did not contain stars and that the Earth was so situated that men could see into the empty region. "Surely," said Herschel, "this is a hole in the heavens."

As more and more of these regions were studied, it began to seem more and more improbable that such regions, unexplainably empty of stars, could exist in such numbers and that all should just happen to be so situated that we could look into the "hole." By 1919, E. E. Barnard had listed the positions of 182 such dark regions, and by now the number of those observed has increased to over 350.

It was borne in on Barnard and on the German astronomer Max Franz Joseph Cornelius Wolf (1863–1932) that these dark regions were not merely holes. They did not indicate the absence of matter but rather the presence of matter, vast clouds of dust particles that absorbed and blocked off the light of the stars that lay behind them, much as the clouds in the Earth's atmosphere absorbed and blocked off the light of the Sun behind them.

In short, there were "dark nebulae" as well as bright ones. The bright ones shone only because they contained stars; the dark nebulae were dark because they did not contain stars.

Famous dark nebulae include the "Horsehead Nebula" in the constellation Orion, which stands out like a dark horse's head against the luminosity of a bright nebula all about. (Actually, the shape of the Horsehead Nebula seems to me to be more like the head and shoulders of Walt Disney's "Big Bad Wolf.") There is also the "Coal Sack," a region of intense darkness near the Southern Cross.

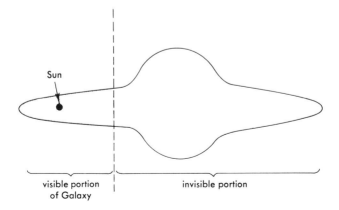

Visible portion of the Galaxy

Sun

visible portion
of Galaxy

invisible portion

If the dark nebulae are more or less evenly distributed throughout the Galaxy, they would be most plentiful exactly where stars are most plentiful. And, indeed, most of the dark nebulae are located in the plane of the Milky Way and, particularly, in the direction of Sagittarius, where the Galactic center and the bulk of the Galaxy's structure are to be found. It is easy to see, then, why the Galactic center and all that lies beyond should be obscured. Their light is permanently blocked off by clouds of dust.

Near the plane of the Milky Way, we also fail to see globular clusters or any very distant objects of types common in other portions of the sky. This is called the "zone of avoidance."

The band of the Milky Way seems roughly equal in brightness in all directions, not because the Solar system happens to be near the center of the Galaxy, but only because most of the light of the Galaxy is obscured by dark nebulae. What we see of the Milky Way is only a portion of our own neighborhood, so to speak—only our own end of the Galaxy.

Thus, Shapley's conclusion that we were eccentrically placed in the Galaxy held fast even though the nature of the visible Milky Way made us seem to occupy a central position.

Galactic Dimensions

Nevertheless, Shapley's conclusion was not correct in detail. For the first time, an astronomer had overestimated the size of a significant portion of the Universe, after thousands of years in which the size of the Universe had been consistently underestimated.

Again, the difficulty arose in connection with what one could not see. Once we grant the existence of dark nebulae it becomes reasonable to wonder whether any part of space is perfectly transparent.

We can draw an analogy here with our atmosphere. We all know that unusual concentrations of dust particles (whether solid as in smokes, or liquid as in fogs, mists, or clouds) can block off light of even something as bright as the Sun. If the dust particles are low-lying (notably as fogs), vision is cut off at moderate distances; in extreme cases, we cannot see more than a few feet, so efficiently do the particles of the fog or smoke scatter light.

It would certainly seem, in comparison, that when the air is sparklingly clear and free of any visible trace of cloud or mist, it is then perfectly transparent and in no way interferes with light.

This, however, is an illusion. Even at its clearest, the atmosphere is not perfectly transparent. It will always scatter light, as is evidenced by the mere existence of a blue sky—the blue being built up of scattered sunlight. And even the clearest night air absorbs a significant fraction of the starlight falling on the Earth.

We can, in short, expect perfect transparency of nothing short of perfect vacuum. Interstellar space (that is, the space between the stars) is nearly a perfect vacuum—but not quite. It is far more transparent, even in the midst of a nebula, than is the Earth's atmosphere, but it is still not perfectly transparent. An occasional particle of dust floating here or there in the enormous emptiness of interstellar space will intercept and deflect a ray of starlight. Taken singly, this amounts to nothing, but in the stretch of light-years between one star and the next, enough dust particles may intervene to result in a cumulative scattering large enough to detect.

This minute scattering effect can best be detected by the fact that short-wave light is more easily scattered than is long-wave light. Such scattering subtracts the blue-violet end of the spectrum and what light remains is increasingly reddened. If, then, the distant stars can be shown to be dimmer and redder than they ought to be, and if this effect increases steadily with distance, then the presence of interstellar dust is strongly indicated.

The first to show definitely that this effect existed was the Swiss-American astronomer Robert Julius Trumpler (1886–1956). He did this in connection with his studies of star-clusters. Such clusters have an average size and an average brightness, and both should decrease in the same manner—in proportion to the square of the distance. It followed then, if a globular cluster takes up a particular area of the sky, it also ought to have a particular brightness.

In 1930, Trumpler showed, however, that the light of the more distant globular clusters was dimmer than was to be expected from their sizes. The more distant the cluster, the more marked this departure from the expected brightness. The area of the clusters was decreasing in accordance with the square of the distance, but the brightness of the clusters seemed to be decreasing in accordance with the square of the distance *plus* some additional dimming effect. And the more distant the cluster, the redder it seemed.

The easiest way of explaining this was to suppose that incredibly thin wisps of dust in interstellar space had a distinctly dimming and reddening effect over vast distances. Over those vast distances, however, the dust succeeded in dimming and reddening the farther clusters, just as the dust in the atmosphere of the Earth dimmed and reddened the setting Sun. In the plane of the Milky Way where dust concentrations are highest, it is estimated that half the energy of a light ray is scattered after a journey of 2000 light-years, half of what is left is scattered in another 2000 light-years, and so on. After 30,000 light-years (the distance of the Galactic center from us) only 1/32,000 of the energy of light rays would be left, even if they did not pass through any of the unusual dust concentrations represented by the dark nebulae. It is no wonder that we cannot see the center of the Galaxy by visible light. (Infrared light penetrates dust more easily and a strong patch of infrared has been detected at a point in the sky corresponding to the invisible center.)

This dimming effect is important in connection with distant Cepheids. From the period of a particular Cepheid, its luminosity can be determined. If this luminosity is compared with its observed brightness, its distance can be determined, since it is easy to calculate how far off a star must be in order for its luminosity to be reduced to the mere pinpoint of light that is actually observed. This calculation assumes, however, that the reduction of brightness is caused entirely by the distance factor. The presence of interstellar dust would perceptibly dim the light of a distant Cepheid by an additional amount, however. If it were not for the presence of dust, such a Cepheid would appear distinctly brighter and would then be judged closer. In other words, the existence of interstellar dust (if not allowed for) tends to produce a falsely large estimate of great distance.

Shapley's estimate that the center of the Galaxy was 50,000 light-years away was based on the assumption that the Cepheids lost brightness only through distance. Allowing for the presence of interstellar dust as a second dimming agency, the center of the Galaxy need be less than 30,000 light-years from us. Even this places us far from the Galactic center, but we remain within 45 light-years of the Galactic plane—the imaginary plane that cuts through the center of the Galaxy lengthwise. That is why the Milky Way seems to divide the sky into two roughly equal halves.

By the early 1930's, then, the dimensions of the Milky Way, as we now accept them, were finally determined. The Galaxy is a lens-shaped object about 80,000 to 100,000 light-years across, with our own Solar system about 27,000 light-years from the center. The Galaxy is about 16,000 light-years thick at the center and 3000 light-years thick at the position of our Sun. The globular clusters are distributed spherically about the center of the Galaxy and that sphere of clusters has an overall diameter of 100,000 light-years.

Furthermore, from the Cepheids observed in the Magellanic Clouds, it was possible to determine their distances. The Large Magellanic Cloud is about 155,000 light-years from us and the Small Magellanic Cloud about 165,000 light-years away. Their position is shown in the facing figure, and there is reason to think that there are tenuous connections between the two Clouds so that they form a single system. Indeed, the Large Magellanic Cloud may have been even closer in the past. About 500 million years ago, it may have brushed our Galaxy, coming within 65,000 light-years of its center.

The size of the Galaxy is not merely a matter of the space it takes up. How many stars does it contain?

To answer that question, Jan Oort assumed that the Milky Way was strongly concentrated toward its center. The central region might contain 90 percent of all the stars in the Galaxy. If this were so, then the stars in the outskirts (our own Sun, for instance) would revolve about the center as, within the Solar system, the Earth revolves about the Sun.

Furthermore, stars closer to the Galactic center than the Sun is would revolve about that center at a greater velocity than the Sun did, while stars farther from the center would move at

The Magellanic Clouds and the Milky Way

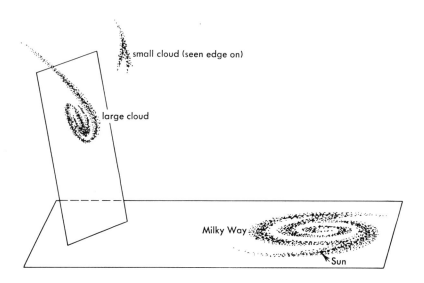

a lesser velocity. (This follows from gravitational theory and is analogous to the manner in which Mercury, for instance, which is closer to the Sun than the Earth is, moves about the Sun at a greater velocity than the Earth does, while Jupiter, farther from the Sun than the Earth is, moves at a lesser velocity.)

This arrangement holds true only if the various bodies moving about the center have orbits that are circular or nearly so, as is the case for the planets in our Solar system. If some stars have very eccentric orbits about the Galactic center, their motions would not be so easily analyzed.

If, however, a large number of stars is taken into consideration, the effects of orbital eccentricity will average out. In that case, it can be seen that the stars that lie between us and the Galactic center will tend to move faster than we do and will catch up to our position, moving slightly toward us, radially, as they do so. Once they pass us, they continue to move farther and farther ahead of us, moving slightly away from us, radially, as they do so. On the other hand, stars farther from the center than we are move more slowly. We catch up to them and move closer to them radially, then pass them and move away from them.

On the whole, then, there should be a certain regularity both in transverse velocities and radial velocities among the stars fairly close to us. And there is. It is the regularity in radial velocities that made it possible to detect the apparent motion of

Galactic rotation

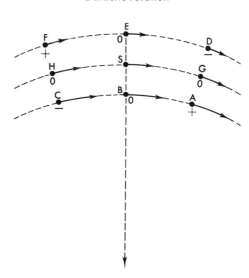

the Sun toward the apex (see page 52), a minor motion, relative to the closer stars only. In 1904, Kapteyn had drawn more general conclusions, deciding that there were two streams of stars, one moving in a particular direction, the other in a directly opposite direction.

In 1925, Oort showed that Kapteyn's streams consisted of the inner stars catching up to the Sun, and the other stars lagging behind the Sun. He was able to determine the nature of the general rotation of the Galaxy and from that calculate the direction and distance of the Galactic center by a method independent of the position of the globular clusters. He showed that the center was some 30,000 light-years distant, in the direction of Sagittarius. This agreed with the evidence of the globular clusters once the presence of interstellar dust was taken into account, and such agreement sufficed to bring Shapley's eccentric model of the Galaxy into general acceptance by astronomers.

Furthermore, Oort could show, from his study of the relative motions of the stars, that the Sun was moving in a fairly circular orbit about the Galactic center at a velocity, relative to that center, of 220 kilometers per second (or about 140 miles per second), completing one revolution about the center in 230,000,000 years.

For gravitational attraction to drive the Sun about its orbit at that distance and that velocity, the Galactic center must have a mass about ninety billion times that of the Sun. If we assume that 90 percent of the mass of the Galaxy is concentrated in its center, then the entire mass of the Galaxy is equal to about 100 billion times that of the Sun.

Statistical studies of the stars seem to indicate that the Sun is an average star in mass, so that we might suspect the Galaxy to contain about 100,000,000,000 stars, although some estimates place the figure higher.

Once allowance is made for their immense distance, the Magellanic Clouds turn out to spread over a region of respectable size. The Large Magellanic Cloud may be some 40,000 light-years in extreme diameter, or nearly half the diameter of the Galaxy. However, the Magellanic Clouds are much less densely populated with stars than the Galaxy is, and that is perhaps the more important fact. The Large Magellanic Cloud contains no more than five to ten billion stars, perhaps, and the Small Magellanic Cloud only one to two billion stars. In terms of star content, the two, taken together, are only one-tenth as large as the Galaxy.

They might also be pictured as "satellites" of the Galaxy.

Nevertheless, despite the smaller size of the Magellanic Clouds, they contain some kinds of objects larger and more spectacular than any found in the Galaxy — at least in those parts of the Galaxy we can see. For instance, the most luminous known star, S Doradus, is in the Large Magellanic Cloud. (It is so-called because the Cloud is in the constellation, Dorado the Swordfish.) S Doradus is not quite bright enough to see with the unaided eye, being only an eighth-magnitude star. To be even so bright, however, at the distance of the Larger Magellanic Cloud, is monumental. S Doradus is thirty times as luminous as Rigel, the most luminous of those stars comparatively near to us and is about 600,000 times as luminous as the Sun.

The Larger Magellanic Cloud also contains the Tarantula Nebula, a bright cloud of dust like the Orion Nebula, but 5,000 times larger. The Tarantula Nebula is far larger and more spectacular than any similar object that can be observed in the Galaxy.

We have, then, a clear picture of that portion of the Universe containing our own gigantic cluster of stars. If we imagine a sphere with its center at the center of the Galaxy and a radius of about 200,000 light-years, it would contain the entire Galaxy and the Magellanic Clouds, including a total of perhaps as much as 150,000,000,000 stars.

CHAPTER **6**

Other Galaxies

The Andromeda Nebula

Man's vision of the size of the Universe had increased enormously in 2000 years. Let us recapitulate.

By 150 B.C., the Earth-Moon system had been accurately defined. The Moon's orbit was seen to be half a million miles across, and the diameter of the planetary orbits was suspected to be in the millions of miles.

By 1800 A.D., the scale of the Solar system had been defined. Its diameter was not merely in the millions of miles, but in the billions. The distance of the stars was still unknown but was suspected to be in the trillions of miles (that is, a couple of light-years) at least.

By 1850 A.D., the distance of the nearer stars had been defined as not merely trillions of miles, but tens and hundreds of trillions of miles. The diameter of the Galaxy was still unknown but was suspected to be in the thousands of light-years.

By 1920 A.D., the diameter of the Galaxy had been defined at not merely thousands of light-years but many tens of thousands of light-years.

At each new stage, the size of the regions of the Universe under investigation turned out to exceed the most optimistic estimates of the past. Furthermore, at each stage, there was the

81

conservative opinion that the object whose size had been defined represented all, or almost all the Universe, and until 1920, that view had always turned out to be wrong.

The Earth-Moon system had shrunk to insignificance in the light of the size of the Solar system. The Solar system had in turn shrunk to insignificance when the distance of the nearby stars was determined. And the system of the nearby stars was insignificant in comparison with the Galaxy as a whole.

Would this process continue or did the Galaxy and its Magellanic satellites represent an end at last? Had astronomers finally probed to the end of the Universe?

Even as late as 1920, it seemed quite possible that the conservative view would finally triumph. The Galaxy and the Magellanic Clouds seemed very likely to contain all the matter in the Universe and beyond them, one could maintain, lay nothing.

This time, there were strong theoretical arguments to back the conservative view. Remember that Olbers' paradox seemed to imply the existence of a finite Universe (see page 44) and the fact that the stars seemed confined to a finite lens-shaped Galaxy bore this out. If there proved to be numerous enormous objects beyond the Galaxy and its satellites, then Olbers' paradox might well present astronomers with an insoluble dilemma.

And yet astronomers could not relax completely with their finite Universe 200,000 light-years across. There were grounds for some suspicion that numerous large objects might exist far outside the Galaxy, and it proved extraordinarily difficult to argue that suspicion out of existence.

A particularly troublesome item was a cloudy patch of light in the constellation Andromeda, an object that was called the "Andromeda Nebula" because of its location and appearance.

The Andromeda Nebula is visible to the naked eye as a small object of the fourth magnitude that looks like a faint, fuzzy star to the unaided eye. Some Arab astronomers had noted it in their star maps, but the first to describe it in modern times was the German astronomer Simon Marius (1570–1624) in 1612. In the next century, Messier included it in his list of fuzzy objects that were not comets. It was thirty-first on his list, so that the Andromeda Nebula is often known as "M31."

There was no reason, at first, for thinking that the Andromeda Nebula was significantly different from other nebulae such as the Orion Nebula. The Andromeda Nebula seemed a luminous cloud and no more than that.

Some eighteenth-century astronomers even envisaged a place for such clouds in the scheme of things. What if stars developed out of distended rotating masses of gas? Under the effect of their own gravity, such clouds would begin to contract and condense, speeding their rotation as they did so. As they rotated more and more quickly, they would flatten into a lens shape and, eventually, eject a ring of gas from the bulging equator. Later, as rotation continued to speed up, a second ring would separate, then a third ring, and so on. Each ring would coalesce into a small planetary body, and finally what was left of the cloud would have condensed into a large glowing star that would find itself at the center of a whole family of planets.

Such a theory would account for the fact that all the planets of the Solar system were situated nearly in a single plane, that all of them revolved about the Sun in the same direction. Each planet, moreover, tended to have a system of satellites that revolved about it in a single plane and in the same direction, as though the planets, in the process of contracting from gaseous rings, gave off smaller rings of their own.

The first to suggest such an origin of the Solar system was the German philosopher Immanuel Kant (1724–1804) in 1755. A half-century later, the French astronomer Pierre Simon de Laplace (1749–1827) published a similar theory (which he arrived at independently) as an appendix to a popular book on astronomy.

It is interesting that Kant and Laplace had opposing views on the Andromeda Nebula, views that kept astronomers at logger-heads for a century and a half.

Laplace pointed to the Andromeda Nebula as possibly representing a planetary system in the process of formation; indeed its structure is such that it seems to be in the obvious process of rapid rotation. You can almost make out (or convince yourself you are making out) a ring of gas about to be given off. For this reason, Laplace's suggestion as to the method of formation of planetary systems is known as the "nebular hypothesis."

If Laplace were correct and if the Andromeda Nebula were a volume of gas serving as precursor for a single planetary system, it cannot be a very large object and, in view of its apparent size in the telescope, it cannot be a very distant object.

Laplace's nebular hypothesis was popular among astronomers throughout the nineteenth century, and his view of the Andromeda Nebula represented a majority opinion through all that time. In 1907 a parallax determination was reported for the Andromeda Nebula, one that seemed to show it to be at a distance of 19 light-years. Certainly, that seemed to settle matters.

Yet there was Kant's opposing view. Despite the fact that he, too, had originated a nebular hypothesis, he did not fall prey to the temptation of accepting the Andromeda Nebula as visible support of his theory. He suggested instead that the Andromeda Nebula, and similar bodies, might represent immensely large conglomerations of stars, which appeared as small, fuzzy patches only because they were immensely far away. He felt they might represent "island universes," each one a separate galaxy, so to speak.

However, this suggestion of Kant's was not based on any observational data available to the astronomers of the time. It made very few converts and, if Kant's speculation was thought of at all, it was dismissed as a kind of science fiction.

But Kant's suggestion did not die. Every once in a while some small piece of evidence would arise that would not quite fit the orthodox Laplacian view. Chief among these was the matter of spectroscopic data.

Stars, generally, produce light which, on passing through a prism, broadens into an essentially continuous spectrum, broken by the presence of dark spectral lines. If, however, gases or vapors of relatively simple chemical composition are heated until they glow, the light they emit, when passed through a prism, produces an "emission spectrum" consisting of individual bright lines. (The exact position of the bright lines depends on the chemical composition of the gas or vapor.)

Then, too, a continuous spectrum usually (but not always) implies white light, while an emission spectrum is often the product of colored light, since some of the bright lines of one particular color or another might dominate the entire glow.

Many bright nebulae do indeed show very delicate color effects (that do not show up in ordinary black-and-white photographs). When Huggins studied the light of the Orion Nebula, for instance, he found that an emission spectrum was produced with a particularly dominating line in the green. One could conclude from this that the Orion Nebula and other objects like it contained masses of hot, glowing gas.[1]

The light from the Andromeda Nebula was a drab white, however, and in 1899 its spectrum was obtained and shown to be continuous.

If the spectrum of the Andromeda Nebula had been shown to consist of bright lines, then the matter would have been settled. It would have been a mass of glowing gas, of no greater significance to the general structure of the Universe than the Orion Nebula was. As it was, the dispute continued. White light and a continuous spectrum meant that the Andromeda Nebula might consist of a mass of stars and be so far off that those stars could not be made out separately. On the other hand, that conclusion was not inevitable, for gaseous nebulae might, under some circumstances, possess white light and continuous spectra.

This was so because emission spectra were produced by hot gases glowing with their own light. Suppose, though, that a mass of gas was cold and was serving merely as a passive reflector of starlight. In that case, the spectrum of the reflected starlight would be essentially the same as the spectrum of the original starlight itself (just as the spectrum of moonlight is like that of sunlight).

If the Andromeda Nebula were merely reflecting starlight, that would explain everything. Its spectrum would be consistent with the theory that it was a not-very-large patch of gas quite close to the Solar system.

But one catch remained. If the Andromeda Nebula were merely reflecting starlight, where were the stars whose light it was reflecting? One could easily see stars within the Orion Nebula, and it was the radiation from these stars that heated the Orion Nebula into a heat great enough to produce an emission spectrum. But where were the stars in the Andromeda Nebula? None could be found.

At least, no permanent stars could be found. Occasionally, a starlike object was found to be associated temporarily with the Andromeda Nebula. As this turned out to be highly significant, let us pause in order to take up the matter of temporary starlike objects in some detail.

[1] Gas is not as efficient as dust particles in scattering light. The fact that nebulae do scatter light efficiently shows that they must be made up of dust as well as gas. As far as we can tell at present, dust makes up only 1 or 2 percent of the total mass of the average nebula, but that is enough to explain the scattering ability.

Novae

To any casual observer of the heavens, the starry configurations seem permanent and fixed. Indeed, the Greek philosophers had differentiated between the sky and the earth by this fact. On Earth, Aristotle suggested, there was perpetual and continuing change, but the heavens were absolutely changeless.

To be sure, there were occasional "shooting stars" which made it appear, to the uninitiated, that a star had fallen from heaven. However, no matter how many shooting stars appeared, no star was ever observed to be missing from its place as a result. Consequently, such shooting stars were considered to be atmospheric phenomena by the Greeks and therefore, like the shifting of clouds or the falling of rain, to be part of the changing earth and not of the changeless heavens. The very word "meteor" applied to shooting stars is from a Greek term meaning "things in the air."[2]

The Greeks were correct in deciding that the flash of light accompanying a shooting star was an atmospheric phenomena. The object causing that flash, though, was a speeding body (a "meteoroid") varying in size from less than a pinpoint to a multi-ton object. Before entering the earth's atmosphere, a meteoroid is an independent body of the Solar system. After entering the atmosphere, it heats through friction to the point where it flashes brilliantly. If small, it is consumed in the process; if large, a remnant may survive to strike the Earth's surface as a "meteorite."

Another class of temporary inhabitants of the sky were the occasional comets, often sporting long, cloudy projections that might be considered as flowing tails or streaming hair. The ancients viewed it as the latter, for "comet" is from the Latin word for "hair." Comets came and went erratically, so the Greek philosophers considered them to be atmospheric phenomena also. Here, they were clearly wrong, for the comets exist far beyond Earth's atmosphere and are actually members of the Solar system, as independent a set of members as the planets themselves.

Nevertheless, suppose we modify the Greek view and say that change is a property of the Solar system, but that the stars far beyond the Solar system are changeless. If we do this, we

---

[2] This explains, by the way, why "meteorology" is the study of the atmosphere and the weather, and not of meteors. The latter study is now termed "meteoritics."

eliminate not only meteors and comets, but also such changes as the phases of the Moon, the spots on the Sun, and the complicated motions of the planets. Is this restricted view of changelessness tenable?

To the naked eye, it would almost seem to be. To be sure the intensity of the light produced by some stars varies (see page 54), but such cases are few and unspectacular, not obvious to the casual eye. Some stars also have significant proper motions, but this is even less noticeable, and it would take many centuries to be sure of the existence of such motions without a telescope.

One type of spectacular change, however, *could* take place in the heavens, and so clearly that the most casual observer could see it. I am referring to the actual appearance of a completely new, and sometimes very bright, star in the sky. Such stars were clearly stars and lacked all trace of the fuzziness of comets. Furthermore, they were not momentary flashes like meteors, but persisted for weeks and months.

Not only were such new stars evidence of change among the stars by the mere fact that they appeared and eventually disappeared, but also they changed brightness radically during the course of their brief stay in the sky as visible objects. Only the fact that such objects were so rarely encountered made it possible for the ancient astronomers to ignore their existence and to continue to accept the assumption of the changelessness of the heavens.

There is evidence, in fact, of only one such new star having appeared during the period of Greek astronomy, and that evidence is none too strong. Hipparchus is supposed to have recorded such a new star in 134 B.C. We do not have his word for this, for virtually none of his works have survived. The Roman encyclopedist Pliny (23–79 A.D.), writing two centuries later, reported it, saying that it was this new star that inspired Hipparchus to prepare the first star map, in order that future new stars might be more easily detected.

Perhaps the most spectacular new star in historic times was not observed in Europe at all, for it appeared in the constellation Taurus in June 1054, at a time when European astronomy was virtually nonexistent. That we know of it at all is thanks to the observations of Chinese and Japanese astronomers who recorded the appearance of what they called a "guest star" at this time. It persisted for two years and grew so fiercely brilliant at its peak as to outshine Venus and become easily visible by day. For almost a month it was the brightest object in the sky next to the Sun and the Moon.

Then, in November, 1572, another such object, almost as bright, appeared in the constellation Cassiopeia, outshining Venus, at its peak, by five or ten times. By then, however, European astronomy was flourishing again, and an astronomer of the first rank was in his impressionable youth. This was the Danish astronomer Tycho Brahe, who observed the new star carefully and then, in 1573, published a small book about it. A short version of the Latin title of the book is *De Nova Stella* ("Concerning the New Star"). Ever since, a star that suddenly appears where none was observed before has been called a "nova" ("new").

One of the important points made in connection with the nova of 1572 by Tycho was that it lacked a measurable parallax. This meant it had to be many times as distant as the Moon and could not be an atmospheric phenomenon and therefore part of the changeable Earth. (Tycho made the same observation in 1577 for a comet and showed that comets, too, were not atmospheric phenomena.)

Then, in 1604, still another nova appeared, this time in the constellation Ophiuchus. It was observed by Kepler and Galileo. While distinctly less bright than Tycho's nova, that of 1604 was still a remarkable phenomenon and, at its peak, rivaled the planet Jupiter in brilliance.

Oddly enough, no superlatively bright novae have graced the sky in the three and a half centuries since 1604. This is rather a pity, for the telescope was invented a few years after 1604, and astronomy entered a new era in which such spectacular novae could have been studied much more profitably than before.

Nevertheless the telescopic revolution in astronomy at once affected the views concerning these novae. In the first place, it was quickly seen that the stars visible to the naked eye were not all the stars there were by any means. A nova, therefore, need not be a truly new star, despite its name. It might merely be a dim star — too dim to be seen by the naked eye, ordinarily — which, for some reason, brightened sufficiently to become visible. As astronomers began to discover more and more variable stars, such changes in brightness came to seem neither phenomenal nor even unusual in themselves. What was unusual about novae was not the fact that their brightness changed, but the *extent* to which it changed. Novae could be classified as a type of variable star, but a particular type called "cataclysmic variables." Their changes in brightness seemed not merely the result of some more or less quiet periodic process, but rather the consequence of some vast cataclysm — somewhat like the difference between a periodic geyser spout and an erratic and unpredictable volcanic eruption.

Then, too, whereas in pretelescopic days, only those sudden brightenings which reached unusual peaks could readily be observed, the telescope made it possible to observe much less drastic events.

Since novae were associated with such brightness, dim ones were not searched for, and for two and a half centuries no novae were reported. Then, in 1848, the English astronomer John Russell Hind (1823–1895) happened to observe a star in Ophiuchus that suddenly brightened. At its brightest, it only reached the fifth magnitude so that it was never anything more than a dim star to the naked eye, and in pretelescope days it might easily have gone unnoticed. Nevertheless, it was a nova.

Thereafter, novae of all brightnesses were searched for and discovered in surprising numbers. One of them, appearing in the constellation of Aquila in 1918 ("Nova Aquilae"), shone, briefly, as brightly as Sirius. None, however, approached the planetary brightness of the novae of 1054, 1572, and 1604.

It is now estimated that some two dozen novae appear each year, here and there in the Galaxy, although relatively few of them are so situated as to be visible from the Earth.

The Andromeda Galaxy

The matter of the novae entered the problem of the Andromeda Nebula when, in 1885, one appeared in the central portions of the nebula. For the first time, a prominent star was seen in connection with the Andromeda Nebula.

There were two possibilities here. The star might exist between the Andromeda Nebula and ourselves and be seen in the nebula only because that object was in the line of sight. In that case the star and the nebula would have no true connection. The second possibility was that the Andromeda Nebula was made up of stars too dim to be seen and that one of them had flared up into a nova and had become visible in a telescope.

If the latter were the case, it might be possible to determine the distance of the Andromeda Nebula if one assumed that novae always reached about the same peak of luminosity. In that case, variations in apparent brightness would be caused entirely by a difference in distance. If the distance of any nova could be determined, the distance of all the rest could then be calculated. The opportunity came with a nova that appeared in the constellation Perseus ("Nova Persei") in 1901. It was an unusually close nova and its distance was estimated by parallax to be about 100 light-years.

The nova that had appeared in the Andromeda Nebula, referred to now as "S Andromedae," reached only the seventh magnitude at its peak (so that it would never have been visible without a telescope) as compared with a magnitude of 0.2 reached by Nova Persei. If the two novae had indeed attained the same luminosity, S Andromedae would have to be some sixteen times as distant as Nova Persei to account for the difference in brightness. It was argued in 1911, then, that the distance of S Andromedae was 1600 light-years.

If S Andromedae were indeed part of the Andromeda Nebula, that meant the nebula, too, was 1600 light-years distant. If S Andromedae were merely in the line of sight of the nebula, the latter would have to be beyond the nova and even more than 1600 light-years from us. In either case, the nebula was at least 800 times as far from us as had been calculated from the apparent parallactic data obtained in 1907. If the nebula were 1600 light-years distant, it had to be quite large to seem as large in our telescopes as it does. It could scarcely represent a single planetary system in the process of formation as Laplace had supposed. Still, one could not yet accept the Kantian view either. Even at 1600 light-years, the Andromeda Nebula had to be merely a feature of the Galaxy.

This line of argument assumed, however, that S Andromedae and Nova Persei actually reached the same luminosity. What if this assumption were not valid? What if S Andromedae were actually much more luminous than Nova Persei ever was? Or much less luminous? How could one tell?

The American astronomer Heber Doust Curtis (1872–1942) believed that the one way of deciding this matter was to search for more novae in the Andromeda Nebula. What could not be judged in the case of one specimen might become clear in the comparative study of many. He therefore tracked down and studied a number of novae in the Andromeda Nebula, and found himself able to make two points.

First, the number of novae located in the nebula was so

high (about a hundred have been detected so far) that there was no possibility that they were not associated with the nebula. To suppose that all those novae just happened to spring up among stars located in the line of sight between ourselves and the nebula was ridiculous. Such a fortuitous concentration of novae was completely unlikely. This further implied that the Andromeda Nebula was not merely a cloud of dust and gas passively reflecting sunlight. It had to consist of numerous stars—a very large number indeed to have so many novae (a very rare type of star) appear among them. That such stars could not be made out even by large telescopes argued that the nebula was at a great distance. Secondly, all the novae observed in the Andromeda Nebula after 1885 were far dimmer than S Andromedae had been. Curtis suggested in 1918 that these other novae should be compared with Nova Persei, and that S Andromeda was an exceptional, extraordinarily bright nova.

If the ordinary novae in the Andromeda Nebula were set equal in luminosity to Nova Persei, then the distance that would account for the unusual dimness of the former would have to be in hundreds of thousands of light-years, at the very least. Such a distance would also account for the fact that the nebula could not be resolved into stars. At such a distance, individual stars were simply too faint to be made out — unless they brightened enormously, nova-fashion.

But if the Andromeda Nebula were indeed at such a distance, it must be far outside the limits of the Galaxy and, to appear as large as it does, it must be a huge conglomeration of a vast number of stars. It was indeed an island universe of the type Kant had once described.

Curtis' conclusion was by no means accepted by other astronomers, and even Shapley was opposed to him.

Entering the lists, however, was the American astronomer Edwin Powell Hubble (1889–1953). It seemed clear to him that the argument from novae would always seem inconclusive since not enough was known about them. If, however, the Andromeda Nebula were actually an island universe, then perhaps a new telescope — more powerful than any available to nineteenth-century astronomers — might settle the issue by revealing the individual stars in the nebula. From the ordinary stars, far less mysterious than the novae, it might be possible to draw firmer conclusions concerning the nebula.

In 1917, a new telescope (the "Hooker telescope" made possible by the donations of John D. Hooker of Los Angeles) had been installed on Mt. Wilson, just northeast of Pasadena. It had a mirror that was an unprecedented 100 inches in diameter, making it by far the most powerful telescope in the world (and it was to remain the most powerful for a generation).

Hubble turned the Mt. Wilson telescope on the Andromeda Nebula and succeeded in making out individual stars on the outskirts. That was the final settlement of one problem: the nebula consisted of stars and not of gas and dust.

By the end of 1923, Hubble was able to identify one of the stars as a variable showing all the characteristics of a Cepheid. He located other Cepheids soon after.

This was exactly what he needed. Shapley had by then worked out the Cepheid yardstick so that the period of variation of the Cepheids in Andromeda could tell Hubble at once the actual luminosity of those stars, provided one could assume that the same laws governing Cepheids in the Galaxy and the Magellanic Clouds also governed them in the Andromeda Nebula.

Once the luminosity of the Cepheids in the Andromeda Nebula was determined, one could then calculate their distance from their apparent brightness, therefore, the distance of the nebula. Hubble calculated this distance to be approximately 800,000 light-years.

By the mid-1920's, then, the matter was settled, and it has not been questioned since. The Andromeda Nebula is not a member of the Galaxy but is located far beyond its bounds. It is a vast and independent conglomeration of stars, an island universe indeed. Kant was right; Laplace was wrong.

Hubble therefore spoke of the Andromeda Nebula as one of a class of "extra-galactic nebulae," to be distinguished from the ordinary "galactic nebulae" such as that in Orion. Shapley, now converted to the new view, felt such terms to be inadequate. The Andromeda Nebula was not to be compared with the Orion Nebula even by terminology, but only with the Galaxy. The Andromeda Nebula was another galaxy in its own right, and Shapley suggested that all such bodies be termed "galaxies."

Today, therefore, we speak of the "Andromeda galaxy." We distinguish our own galaxy either by giving it a definite article and a capital, "the Galaxy," as I have been doing in the last few chapters, or by calling it "the Milky Way galaxy."

The Spiral Galaxies

Nor was the Andromeda galaxy a sport or a unique example

of something beyond the Galaxy. It was one of a large group, although, to be sure, it was by far the largest in appearance, and,
aside from the Magellanic Clouds, the only one visible to the
naked eye.[3]

Messier, in his 1781 catalogue, listed some dozens of nebulae
which, like the Andromeda, could neither be resolved into many
faint stars nor found to contain a few bright ones, and which, eventually, turned out to be galaxies. William Herschel, in his general
sweeping of the heavens, located no less than 2500 such nebulae,
and his son, John Herschel, in a similar investigation of the skies
of the southern hemisphere found an equal number there. They
were found all over the sky except in the plane of the Milky Way,
and there, as astronomers came to realize, those that might be
present were obscured by the dust clouds and star masses associated with that plane.

By the beginning of the twentieth century, some 13,000 nebulae of the Andromeda type were known, and there was every
indication that many more remained to be found. Today we know
that 50,000 such nebulae exist brighter than the 15th magnitude
and many millions fainter than that.

The Irish astronomer, William Parsons, 3rd Earl of Rosse
(1800–1867), studied these nebulae more closely than did anyone else in the nineteenth century. He had built a 72-inch telescope on his estate, although the weather was usually so poor that
he got very little use out of it. Nevertheless, he studied the nebulae
and noted in 1845 that some of them seemed to have a distinctly
spiral structure, almost as though they were whirlpools of light set
against the black background of space.

The structures appear flat and, when seen edge-on, seem to
be merely elongated lens-shaped objects (like our own Galaxy)
with the spiral structure, if any, invisible. On the other hand, some
nebulae are seen squarely broadside on so that the spiral structure
is fully visible. A particularly spectacular example is "M51," more
dramatically called the "Whirlpool Nebula" (or, today, the "Whirlpool galaxy") from its shape. It was the first to be noted as spiral
by Lord Rosse. The Andromeda galaxy was found to be spiral in
shape in 1888, thanks to the photographs taken by the English
amateur photographer, Isaac Roberts.

As a result, astronomers began to speak of a new class of objects, the "spiral nebulae" or, as they are now called, the "spiral

---

[3] The Andromeda galaxy is, indeed, notable for being the very farthest
object that can be made out by the naked eye.

galaxies." They consist of a central condensation, relatively small in some cases, much larger in other, the "galactic nucleus." Outside are the "spiral arms."

There seems to be a distinct difference in properties between the galactic nuclei and the spiral arms. The nuclei resemble huge globular clusters and, like the clusters, seem relatively free of dust clouds. The spiral arms, on the other hand, are rich in dust clouds, which are often clearly visible.

The dust is most prominent in the case of some spiral galaxies that we happen to view edge-on. An example is NGC 891[4] in the constellation Andromeda. The dust clouds along its equator make a dark, ragged line down its length. This is also true of a beautiful galaxy in the constellation Virgo, in which the galactic nucleus is enormous and the spiral arms rather compact. It is seen almost edge-on, and the dust in the arm forms a tight ellipse around the rim. The effect is almost that of seeing Saturn with dark rings. The dark rim reminds one of a decorated hat brim, and the object is familiarly called the "Sombrero galaxy."

By 1900, the American astronomer, James Edward Keefer (1857–1900) was able to show that about 75 percent of the galaxies have a spiral structure. This, as stated above, includes the Andromeda galaxy, although we see that object so nearly edge-on that the spiral arms are not as clearly evident as we would like. Still, the appearance is striking enough to give the Andromeda galaxy a swirling appearance and make Laplace's suggestion that it is a collection of rotating gas all too plausible.

It is usually considered that our own Galaxy is also spiral in nature, something first suggested by the American astronomer, Stephen Alexander (1806–1883) in 1852. The Galaxy may, indeed, be very much like the Andromeda galaxy in appearance (although recent data seem to indicate that the Galaxy possesses a less prominent nucleus). The Sun is located in one of the spiral arms of the Galaxy, and it is for that reason that we are surrounded by dust clouds that obscure the main body of the Galaxy.

About 20 percent of the galaxies are spheroidal or ellipsoidal and seem to be composed of galactic nuclei only, without spiral arms. These are generally referred to as "elliptical galaxies."

The remaining 5 per cent are "irregular galaxies" with no well-defined simple structure at all. The Magellanic Clouds are

[4]   Galaxies are often referred to by their number in the listing in the *New General Catalogue of Nebulae and Clusters,* published from 1888 to 1908 and usually referred to by the abbreviation, NGC.

often considered to be the best-known examples, but this view was challenged in the 1950's by the French-American astronomer Gerard Henri de Vaucouleurs (1918–   ), who maintained they were spiral in nature. The Large Magellanic Cloud, he pointed out, has a single arm extending outward for some tens of thousands of light-years. The Small Magellanic Cloud is seen virtually edge-on, so that its spiral structure, if present, cannot be made out.

Modern astronomers can set no limit to the number of galaxies. Better astronomic tools and more careful observations continually raise the number that can be seen, and there seems, as yet, no sign of any thinning out in any direction. Astronomers suspect that the total number of galaxies in the parts of the Universe we can observe with our best instruments may be as high as 100,000,000,000, and there is, at present, no reason to think that the actual number may not be indefinitely higher.

Furthermore, the Andromeda galaxy is the brightest member of the class of undoubted spirals and is the nearest large one. If it is 750,000 light-years away, then the dimmest galaxies that can be made out must be many hundreds of millions of light-years away, possibly even billions of light-years.

By 1925, man's notion of the Universe had received another colossal enlargement. Indeed, astronomers found themselves facing the problem of infinity once again. A century earlier, Olbers' paradox had seemed to argue against the possibility of an infinite Universe, and Herschel's observation of a finite Galaxy had tended to bear that out. For a century the notion of a finite Universe had reigned supreme.

But now there was no sign of finiteness in the new and larger Universe of the galaxies, and once again the astronomers would have to tackle the problem of Olbers' paradox, with galaxies taking the place of stars.

This time, however, the problem of the extension of the Universe in space—whether it was infinite or not—turned out to be intimately connected with the allied question of the extension of the Universe in time—whether it was eternal or not.

Until now, I have been considering only the problem of extension in space. Before continuing further in that direction, it will be useful to turn to the problem of extension in time.

CHAPTER 7

The Age of the Earth

Angular Momentum

As long as astronomers considered the heavens to be change-less, the implication was that they were eternal as well. Certainly, a beginning and an ending are the most drastic of all possible changes, and what could be observed of the heavens in ancient times gave no evidence whatever of any possible beginning or ending. (Men might speak of the creation or destruction of the Universe through some superhuman agency and go on to describe the processes in detail, but such descriptions arise from internal inspiration and not from any actual astronomic evidence.)

By early modern times, however, it was recognized that the heavens were not absolutely changeless; the novae were the best evidence of that. The question therefore arose as to whether the possibility of change implied the ultimate change of a beginning and an ending and, if so, when the Universe might have begun and when it might end.

It was easiest to tackle the problem first in connection with the Solar system, which by 1700 was already understood in detail.

In 1687, Newton had established the theory of universal gravitation, according to which every body in the Universe attracted every other body with a force that was proportional to the product of their masses and inversely proportional to the square of the distance between them.

In the Solar system, the Sun was so overwhelmingly predominant in mass that it remained almost motionless, while the much less massive planets responded to the force of Solar gravitation by circling the Sun in elliptical orbits. (Actually, the Sun also moves in response to the gravitational pull of the planets. The center of gravity of the Solar system about which the Sun and planets move is near the center of the Sun but not quite at it. Indeed, it is sometimes so far away from the Sun's center as to be a little outside the Solar sphere. This, however, is a mere detail in a larger picture, and the planets can be viewed as revolving about an essentially motionless Sun without too great an imprecision.)

The Solar system cannot be looked on as changeless, in the sense of being fixed and immobile, because all its component parts (even the Sun itself) are constantly moving with respect to the center of gravity of the system.

But if there is not a static equilibrium, there may, at least, be a dynamic one. That is, although all the parts of the system move, it may be that all the motions are periodic, repeating themselves over and over, endlessly and without significant change, and in that respect the Solar system might be considered to be changeless.

Or is it safe to assume that the motions are truly periodic? To be sure, the Earth would revolve about the Sun in an absolutely periodic manner, never changing its orbit, if the Earth and Sun were the only objects in the Universe. But they are not; there are other objects and each of these affects the Earth-Sun system gravitationally. Neighboring planets affect the Earth's motion through their gravitational influence; so does the Moon; so do even the distant stars.

These minor effects on the Earth's motion ("perturbations") must be taken into account in refined calculations of the Earth's orbit. Perturbations also affect the motions of the other planets.

In the short run, these perturbations do not seriously affect the Solar system. Throughout man's history, the day and the year have remained essentially unchanged, and the motions of the planets have persisted with a grand steadiness. Still, the history of astronomic observations is, at best, only a few thousand years old, and this could be considered a mere moment in the history of the Solar system. What about the long run?

Theoretically, the law of universal gravitation could be applied to predict the motions of every object in the Universe under the gravitational influence of every other body. The machinery of the Solar system could then be run (mathematically) forward and backward in time through indefinitely long stretches—as is literally done, for short stretches of time, in modern planetariums. In this way, one could check as to whether there were any systematic changes that might have pulled the bodies of the Solar system together in the far past or that might force them apart in the distant future.

Unfortunately, such a direct study is not practical. The equations one must set up to account for the motions of only three bodies, gravitationally interlocked, are too complicated for complete solution. What, then, can be done for a Solar system consisting of a dozen major bodies and uncounted numbers of minor ones? Short-cut approximations had therefore to be made and even they required the full-time attention of first-class minds.

The problem was tackled by the French astronomer Joseph Louis Lagrange (1736–1813) and was then followed up by Laplace. It was Laplace who finally solved the matter satisfactorily in his book *Celestial Mechanics,* published in five volumes from 1799 to 1825. He showed that while perturbations introduced small changes into planetary orbits, these changes were periodic; that is, the orbit would alter its properties in one direction, then back in the other, and so on indefinitely. Over the long run, the average shape of the orbit would remain constant.

In other words, the Solar system was in dynamic equilibrium and could continue indefinitely into the future and might already have existed through an indefinite past. (This is predicated on the assumption that there is no introduction of overriding influences from without the Solar system, that no star invades our immediate neighborhood, that the stars as presently situated exert gravitational influences too small to matter, and so on. This is a pretty safe assumption even over long periods of time.)

Although there is nothing in the sheer gravitational mechanics of the Solar system to prevent it from being eternal, the concept of eternity remains a difficult one, no more acceptable to the average mind than that of infinity. For that reason, beginnings were sought by adding something else to gravity.

For instance, if the gravitational attraction of the Sun were the *only* source of movement in the Solar system, the various planets would respond by falling directly toward and into the Sun. The fact that they revolved about the Sun, as did the comets and asteroids, while the various satellites revolved similarly about their planets, meant that each body possessed a motion more or less at right angles to the pull of the Sun's gravitation. This motion did not and could not originate from the Sun's gravitational attraction. How then did it come to be?

This can be put in another way. Any essentially circular motion, whether that of an object rotating about its own axis, or of an object revolving, as a whole, about a larger body, involves the possession, by the moving body, of a property called "angular momentum." The quantity of angular momentum possessed by a body depends on three things: its mass, the speed of its circular motion, and the distance of the body (or the various parts of it) from the center about which it circles.

By the eighteenth century it was quite clear to physicists, through observations of phenomena on Earth, that angular momentum was neither created nor destroyed but that it could be transferred, without loss or gain, from one body to another. This is the "law of conservation of angular momentum." There was (and is today) no reason to think that angular momentum is not conserved in the Universe at large as it is on the Earth. In that case, any theory that describes the beginning of the Universe or any large part of it must not involve any violation of this law.

If angular momentum cannot be created, how does it come to exist? One way out of the apparent dilemma involves the recognition that angular momentum can exist in two varieties according to the direction of spin. There can be "clockwise" angular momentum and "counterclockwise" angular momentum, these varieties being related to the conventional direction of turning of a clock's hands.

If clockwise angular momentum is considered positive, then counterclockwise angular momentum can be considered negative. Equal quantities of the two varieties can, in adding together, cancel, and form a combined system with no angular momentum. Similarly, a system without angular momentum can split into two systems, one of which has a certain clockwise angular momentum and the other an equal counterclockwise angular momentum. In this way angular momentum seems to be destroyed and created without, however, actually breaking the law of conservation of angular momentum.

One might suppose, for instance, that at the beginning the Universe did not contain angular momentum, but that in the process of forming, some portions obtained one variety and the rest the second variety.

Luminous Nebula. *(Lick Observatory Photograph.)*

The Horsehead Nebula. *(Photograph from the Mount Wilson and Palomar Observatories.)*

The Andromeda Nebula. *(Lick Observatory Photograph.)*

Resolution of Andromeda Nebula. *(Photograph from the Mount Wilson and Palomar Observatories.)*

Sombrero Galaxy. *(Photograph from the Mount Wilson and Palomar Observatories.)*

Whirlpool Galaxy. *(Photograph from the Mount Wilson and Palomar Observatories.)*

If the Solar system is viewed from a position high above the Earth's north pole, then the Sun and Earth and most of the other bodies of the system will be viewed as rotating about their axes in a counterclockwise direction. The planets and satellites will almost all be seen to be revolving about their central bodies in a counterclockwise direction. This means that the Solar system does *not* have equal quantities of the two kinds of angular momentum and *cannot* be viewed as a system essentially without angular momentum. Rather, the Solar system has a great deal of angular momentum, and any theory that explains the origin of the system must take that into account.

Suppose, for instance, that the Solar system began as a vast, thin cloud of dust and gas, in accordance with the suggestions of Laplace when he advanced his nebular hypothesis (see page 83). This cloud might already possess a supply of angular momentum, a supply it had received for its share when the Universe as a whole came into being. Or else, if the cloud is supposed to have no angular momentum to begin with, it might come under the influence of the very feeble gravitational attraction of some comparatively nearby star. This would tend to pull more strongly at the end of the cloud nearer itself than at the opposite end. This would exert a "torque" on the cloud and pull it into circular motion. The angular momentum would have been supplied the cloud at the expense of the quantity originally possessed by the attracting star, of course, and the star's own supply would be correspondingly diminished.

Whatever the source of the spin, the slowly rotating cloud would be under the influence of the mutual gravitational attraction of its constituent particles and would slowly contract. As the cloud contracted, its various portions would be closer and closer to the center about which all were revolving. As the overall distance to the center decreased, this, taken by itself, would tend to destroy some of the angular momentum, unless the decrease were balanced by an increase in the velocity of turning. (Angular momentum depends on both factors and on mass in addition, but mass is not changed under conditions described, so only distance and angular speed need be considered. The decrease of either automatically implies an increase in the other.)

The law of conservation of angular momentum therefore makes it necessary for the vast rotating spheroid of gas to spin faster and faster as it contracts. The equator bulges under the influence of a steadily increasing centrifugal effect, turning the spheroid into a more and more flattened ellipsoid. Finally, portions split off the equatorial plane of the ellipsoid at different intervals and condense into planets.

Laplace's nebular hypothesis seemed to explain a great deal, and it bore the appearance of the great plausibility. It was very popular with astronomers throughout the nineteenth century and with the general public as well. Indeed, as the decades of the nineteenth century passed, the nebular hypothesis seemed to fit in very well with certain key discoveries in physics and to offer a method for determining the age of the Earth.

The Conservation of Energy

In the 1840's, a new and even more powerful conservation law than that governing angular momentum came to be established quite firmly. This was the "law of conservation of energy"; it was the product of the work of many men, but it was first clearly enunciated by the German physicist Hermann Ludwig Ferdinand von Helmholtz (1821–1894) in 1847.

The law of conservation of energy states that energy can be transferred from one place to another, but cannot be created out of nothing or destroyed.

The Solar system has an enormous store of energy as well as of angular momentum, and the question must arise as to where that energy comes from.

The nebular hypothesis

There is a vital difference, here, between the question of the Solar system's energy supply and its supply of angular momentum. As far as angular momentum is concerned, one can say that the Solar system's supply was brought into existence when the Solar system was formed (as described in the preceding section), and once that is stated, one can relax. No significant quantity of the angular momentum of the Solar system is being lost. It can be lost only by interaction with the enormously distant stars or with the incredibly thin wisps of matter between the stars. Such processes would take away from, or add to, the Solar system's supply of angular momentum at so slow a rate compared with the total supply present that the whole process could be ignored even over long stretches of time. It seems perfectly safe to assume that the Solar system possesses as much angular momentum now as it did, say, millions of years ago, or as it will possess millions of years from now.

Not so with energy. Energy exists in the Solar system in a variety of forms, and one of these forms is that of the radiation that constantly streams out of the Sun. The amount of energy represented by this radiation is simply colossal, and virtually all of it pours outward in all directions, at an enormous rate, into the vast spaces beyond the Solar system. Moreover, as nearly as we can tell, virtually none of it ever returns.

This means that the total quantity of energy in the Solar system must constantly be decreasing. It must someday decrease to a point so close to zero that the Solar system, as we know it, may be considered at an end. If we look backward in time, the amount of energy in the Solar system is greater and greater the farther back we go. Unless we are content to postulate an infinite quantity of energy to begin with (which no astronomer is), there must be some definite point at which the Solar system began with some original store of energy much greater than that which it possesses today.

Helmholtz himself was the first to worry about this matter and to question the source of the energy that the Sun so recklessly spilled out into space in quantities so magnificent that the utterly contemptible portion intercepted by the tiny Earth at a distance of 93,000,000 miles was sufficient to supply all the energy needs of man with copious excess.

The most common source of man-made energy in the nineteenth century was that of burning coal. In this case, heat and light were obtained at the expense of the chemical energy binding atoms together. When coal and oxygen combined, the carbon dioxide that formed required smaller quantities of chemical energy to keep its molecules in being than was required by the original coal and oxygen. The excess energy no longer required for atomic bonding was then expelled as heat and light.

But the amount of energy that could be produced by combining a given quantity of coal and oxygen was known, and the amount of energy put out by the Sun each second was also known. It is not difficult to show that if the entire mass of the Sun (also a known quantity) were made up of coal and oxygen in correct proportions, the resultant coal fire would keep the Sun going at its present rate for only 1500 years.

Although there are some chemical reactions that yield more energy per pound than a coal fire does, there is not one known that would keep the Sun going throughout historic times, let alone during the long eons of prehistoric times. Helmholtz had to search for his energy source elsewhere.

One colossal source of energy was the gravitational field itself. A meteorite striking the Earth's atmosphere, while moving at the high speed enforced by its response to the gravitational fields about it, converted the energy of motion into light and heat. Even a pinhead meteorite was sufficient to produce a brilliance that could be seen for a hundred miles or more.

Suppose, then, that meteors were constantly plunging into the Sun and that the energy of their gravitationally induced motion were converted into radiation. It can be shown that if such meteorite collisions took place at the rate necessary to keep the Sun going, the Sun would not undergo any visible change in the course of historic times. Indeed, such bombardment could continue for 300,000 years before the Sun would gain 1 percent in mass as a result.

This sounds hopeful but additional thought spoils the picture. The mass of the Sun would increase very slowly as a result of the quantity of meteorite bombardment required to keep it going, but that increase would be enough to introduce a perceptible strengthening of its gravitational field. The Earth, more powerfully attracted each year, would move more and more rapidly, so that each year would be two seconds shorter than the year before. This does not sound like much, but astronomers could detect such a change in the year very easily and virtually at once. Since no such change is detected, the meteor infall theory must be abandoned.

Helmholtz turned, in 1853, to another alternative. What if

the Sun itself were contracting so that its own outer layers were, so to speak, falling toward the center. The energy of this gravitationally induced motion could be converted into radiation just as that of meteors could be, and without any overall change in the Sun's mass either, so that the Earth's year could be left unchanged.

The question was: How much contraction would have to be postulated in order to keep the Sun's radiation going at the necessary rate? The answer was: Very little.

It could be shown that in all the 6000 years of man's civilized history, the Sun's diameter would have contracted by only 560 miles which, in a total diameter of 864,000 miles, can certainly be considered insignificant. The shrinkage in diameter of the Sun over the 250 years from the invention of the telescope to Helmholtz's time would be only 23 miles, a quantity that would be undetectable even with the most refined astronomical instruments of the time.

The source of the Sun's energy was thus explained in terms that seemed satisfactory at the time. In fact, Helmholtz's contraction hypothesis could be combined with Laplace's nebular hypothesis, and one could then envisage energy as having been produced constantly all the time Laplace's original nebula was contracting. The Sun's present-day contraction would only be a final phase of the general nebular contraction.

Moreover, if one assumes that energy was being produced by this contraction all along, at the same rate at which it is now being produced, then one can calculate when the original nebula had attained any particular degree of condensation on its way toward the formation of the comparatively small, hotly glowing Sun that exists today.

For instance, 18,000,000 years ago, the original nebula would have contracted to a diameter of some 200,000,000 miles, and its still bloated sphere would fill all the space out to the present orbit of the Earth. It would have to be 18,000,000 years ago that the ring of matter (according to Laplace's view), which would eventually condense to form the Earth itself, was liberated. The Earth could not, in consequence, be older than 18,000,000 years.

By this line of argument, the planets closer to the Sun than Earth—Venus and Mercury—would have been formed considerably less than 18,000,000 years ago, while the outer planets, Mars, Jupiter, and so on, would have been formed earlier. The entire lifetime of the Solar system, from the beginnings of the nebular contraction, might be several hundred millions of years.

Nuclear Energy

Had Helmholtz presented his theory in 1803 rather than in 1853, the time allotment of 18,000,000 years for the Earth's existence would have seemed satisfactorily long, even excessively so. Indeed, as the nineteenth century opened, most European scientists were still under the spell of the literal language of the Bible and assumed that the Earth had existed for only 6000 years or so. Eighteen million years would have seemed a blasphemously large figure to most of them.

But the first half of the nineteenth century had seen an important revolution in attitude. In 1785, the Scottish geologist James Hutton (1726–1797) had published a book entitled *Theory of the Earth* in which he studied the slow changes that the Earth's surface underwent—the layering of sediment, the erosion of rocks, and so on. He suggested the "uniformitarian principle" which held that whatever changes were going on today had been going on at essentially the same rate throughout the past. According to this principle, it would take enormous stretches of time to produce all the thicknesses of sediments that could be found, all the erosion that could be observed, all the buckling and other forced changes to which the Earth's surface had been subjected.

Hutton did not persuade his readers at the time, but between 1830 and 1833, another Scottish geologist, Charles Lyell (1797–1875), published *The Principles of Geology*. In this book, Hutton's work was summarized, popularized, and backed by additional evidence. This eventually turned the trick, and geologists began to interpret the Earth's history in terms of hundreds of millions of years.

When Helmholtz emerged with his figure of 18,000,000 years as the extreme age of the Earth, geologists were astonished. For a ring of dust and gas to appear 18,000,000 years ago, slowly condense and undergo the changes required to form a compact solid body, with an ocean and atmosphere, and then proceed to undergo all the further changes, after solidification, for which the Earth's crust gave evidence, seemed simply impossible.

Furthermore, biologists were coming to the conclusion that life forms had been slowly changing over the course of time. In 1859, the English naturalist Charles Robert Darwin (1809–1882) published *The Origin of Species* in which he argued that such changes had been brought about by the pressures of natural selection, a process that was excessively slow and required eons of time to produce the changes observed in fossilized record of extinct forms of life.

Darwin's views won out over the Bible-centered prejudices of the time only with great difficulty, but more and more biologists came to accept them and they, too, found they could not swallow Helmholtz's figure. Yet there seemed no disputing Helmholtz's logic, and no quarreling with the law of conservation of energy.

The last half of the nineteenth century witnessed a standoff, then, on the question of the time of origin of the Solar system, and of the Earth in particular. Physicists supported a short lifetime, geologists and biologists supported a long one.

The standoff was broken in the 1890's, when the science of physics underwent a revolution. In 1896, the French physicist Antoine Henri Becquerel (1852–1908) discovered that uranium compounds served as a continuing source of high-energy radiations. (This phenomenon came to be called "radioactivity.") Apparently there were sources of energy much more intense than those involved in chemical reactions or even in gravitational contraction.

By 1911, the New Zealand-born British physicist Ernest Rutherford (1871–1937) had succeeded in demonstrating that the atom was not a featureless sphere but consisted of a tiny "atomic nucleus" at the center, which contained virtually all the mass of the atom and which was surrounded by light particles called "electrons." Chemical reactions involved the forces holding electrons in place around the nucleus, and this was the source of energy for phenomena such as the burning of coal.

The atomic nucleus was itself composed of particles, which were eventually discovered to be of two varieties, "protons" and "neutrons." They were held together in the nucleus by forces many times stronger than those that held the electrons to the nucleus or that held different atoms or molecules together. There are "nuclear reactions" that involve shifts in proton-neutron combinations and that yield much greater intensities of energies than any chemical reaction could. Radioactivity is a form of nuclear reaction.

One aspect of nuclear reactions was brought rather surprisingly to the fore in 1905 by the German-Swiss physicist Albert Einstein (1879–1955). He showed that mass itself was a very concentrated form of energy, and he presented the now well-known formula: $e = mc^2$, where e represents energy, m represents mass, and c represents the velocity of light in a vacuum.

If we remember that the value of c is very high (300,000,000 meters per second) and that the value of c^2 is this tremendous quantity multiplied by itself to produce something vastly more tremendous, we see that even a small quantity of mass is equivalent to a large quantity of energy. Thus, 1 gram of mass can be converted into 21,500,000,000 kilocalories, a quantity that could also be obtained by the complete burning of 670,000 gallons of gasoline.

Release of energy is always at the expense of disappearance of mass, but in ordinary chemical reactions, energy is released in such low quantities that the mass-loss is insignificant. As I have just said, 670,000 gallons of gasoline must be burned to bring about the loss of 1 gram (1/27 of an ounce). Nuclear reactions produce energies of much greater quantities, and here the loss of mass becomes large enough to be significant.

Suppose, for instance, that the Sun received its energy not at the cost of gravitationally induced contraction, but as the result of some nuclear reaction proceeding within it. How much mass would it have to convert into energy in order to radiate energy at its observed rate? This can easily be calculated; it turns out to be 4,600,000 tons of mass per second. This mass would be permanently lost to the Sun, for the energy into which it is converted would be radiated out into interstellar space.

Is it possible for the Sun to support this steady drain of mass at the rate of millions of tons per second? Yes, it certainly is, for the loss is infinitesimally small compared with the total vast mass of the Sun and the trillions of years would have to pass before the loss at such a rate could consume even 1 percent of the mass of the Sun.

Nor would the loss of mass seriously affect the nature of the Earth's gravitational field. The loss proceeds at a rate of only a thirty-millionth that of the gain produced by the meteor-infall theory. The mass loss produced by the nuclear reaction theory would weaken the Sun's gravitational field only to the point at which the Earth's year would increase in length by only one second in 15,000,000 years. Such an increase in length is insignificant.

The nuclear reaction theory requires no perceptible change in the volume or appearance of the Sun over an extended period of time, and it might have existed in very much its present form (and the Earth, too, therefore) not only for tens of millions of years, but for billions of years. The geologists and biologists were vindicated, and Helmholtz's short-lifetime suggestion went by the boards.

Radioactivity itself offered a new method for determining the age of the Earth, and one that was more accurate and reliable than anything previously known.

As uranium gives off its radiations, its atoms change their nature, becoming other kinds of atoms which also give off radiations and change nature again. Eventually, the uranium is converted into lead, which is stable and changes no further.

The rate at which uranium changes in this manner follows a simple rule, well known to chemists as a "first-order reaction." This means that if the rate of change is determined over a short interval of time, it can be predicted, quite accurately, over any longer interval. It could be shown, for instance, that half of any quantity of uranium would break down and change to lead in 4,500,000,000 years. This tremendous time interval is called the "half-life" of uranium-238 (the most common form of the uranium atom).

Suppose, now, that you consider a rock containing uranium compounds. Inside it, the uranium is constantly breaking down and turning into lead. If the rock remains solid and unbroken, the lead atoms formed cannot possibly escape but must remain intermingled with the uranium. The uranium compounds may have been pure to begin with, but they become increasingly contaminated by lead. Since the rate at which nuclear reactions proceed is not affected by the puny changes in temperature and pressure encountered on the Earth, we know that the exact quantity of lead accompanying the uranium depends only on the length of time the rock has remained solid (and on the quantity of lead present originally) and not on any unpredictable environmental changes to which it may have been subjected.

This was pointed out as early as 1907 by the American chemist Bertram Borden Boltwood (1870–1927); in the years following, rocks were subjected to analysis for uranium and lead content, and techniques were worked out for converting these analyses into age measurements. Within a few years, rocks were discovered that, by the uranium-lead method, must surely have been lying in an undisturbed solid state for periods in excess of a billion years.

In the last few decades, a variety of age-determining methods based on one form of radioactive change or another have yielded reliable value of 4,700,000,000 years for the age of the Earth—a period 260 times as long as that suggested by Helmholtz.

8

The Energy of the Sun

The Planetesimal Hypothesis

With the great multibillion-year age of the Earth well-established by 1920, it was natural to ask what the age of the Sun might be. If the nebular hypothesis were a true picture of the development of the Solar system, it would follow that the oldest planet was the outermost, the youngest the innermost, and the Sun in its present form younger than any planet. If the age of the Earth is taken as 4.7 eons,[1] it would follow that the Sun must be somewhat younger than 4.7 eons, but not, perhaps, much younger than that.

Unfortunately, no such easy conclusion could be reached, for the nebular hypothesis, which had remained generally popular throughout the nineteenth century, had gone out of fashion at the turn of the century.

The problem over which the nebular hypothesis stumbled and fell was angular momentum. The nebular hypothesis began with a vast quantity of dust and gas that contained a supply of angular momentum. It visualized the process of condensation with its concomitant steady increase in the rate of rotation of the cloud and the eventual splitting off of successive shells of dust and gas. It

---

[1] It is becoming increasingly common to make use of the word "eon" (ordinarily used for any indefinitely long period of time) to stand for a billion years.

111

made no attempt, however, to describe how the angular momentum was to be divided between the shells that split off and formed the planets and the main portion of the cloud that continued to condense to form the Sun.

In 1900, the American geologist Thomas Chrowder Chamberlin (1843–1928) worked out the dynamics of a spinning nebula very carefully. He showed that as the nebula gave up a shell of equatorial matter and continued to contract, virtually all the angular momentum would have to remain with the main body of the nebula and very little could be left with the shell. If such a shell could then coalesce into a planet (a very doubtful process, as it turned out), the planet would have very little angular momentum. The final result would be a Solar system in which the central Sun would contain almost all the angular momentum of the entire system and would therefore be turning on its axis very rapidly, with a period of about half a day, in fact. The planets would contain so little angular momentum that it was doubtful whether they could stay in any reasonable orbit at all.

But this is *not* the picture of the Solar system as it actually exists. In actual fact, the single planet Jupiter, with only about 0.2 percent of the mass of the Solar system, contains fully 60 percent of the total angular momentum in the system. Saturn contains another 25 percent.

Despite the fact that Jupiter has eleven times the diameter of the Earth, the period of rotation of the larger planet is 10 hours, less than half the period of Earth's rotation. Saturn, almost as large as Jupiter, rotates almost as quickly. Both massive worlds swing in great arcs about the Sun, and it is this revolution that contains most of the angular momentum.

Add the other planets and minor bodies of the system (all of which taken together contain less than 0.1 percent of the mass of the system) and the total planetary angular momentum is found to be 98 percent of the system-wide total. The Sun with over 99.8 percent of all the mass of the Solar system contains only 2 percent of all its angular momentum and rotates on its axis with majestic deliberation completing a turn only after 24.65 days.[2]

How could a nebula contract and in doing so, shift almost all its angular momentum to the tiny shells of matter it gave

[2]   This is the period of rotation at the Sun's equator. The Sun is not a solid body and does not rotate all in one piece as does the Earth. Points north and south of the equator rotate with a longer period. At 60° north or south latitude, the period is about thirty-one days.

off? Chamberlin could find no plausible way to account for this. He had to conclude that the angular momentum was brought into the Solar system from outside.

In 1906, Chamberlin, together with the American astronomer Forest Ray Moulton (1872–1952), suggested a way out. Picture the Sun, to begin with, in much its present form, but without planets. Perhaps it had condensed from a nebula to begin with, but if so, it condensed without liberating shells of matter; or, if shells were liberated, they lacked sufficient angular momentum to remain independent but gradually fell back into the main body or drifted off into space. In any case, the Sun exists in lonely splendor.

Picture a second star approaching the Sun. The tresmendous gravitational forces that result would produce huge tides on both stars. Perhaps a gigantic gout of star-matter might rise out of both stars and form a temporary bridge between them. As the stars passed each other, this bridge of matter would be forced to swing rapidly round and would gain angular momentum at the expense of the stars themselves.

Once the stars separated, each would carry off some share of the matter-bridge, which would then coalesce into planets, still retaining the angular momentum they had gained. Before the approach the two stars would have been fast-spinning with no planets; after the approach they would be slow-spinning with orbiting planets.

The objections to the nebular hypothesis seemed conclusive, and the Chamberlin-Moulton theory seemed a neat substitute. In fact, it seemed particularly attractive since it allowed what was almost a biological motif to enter astronomy. It was as though planets were formed by a kind of marriage between two stars, as though the Earth had both a father and mother. This hypothesis was undisputed for nearly forty years.

Because Chamberlin and Moulton pictured the matter pulled out of the Sun as quickly condensing into small solid bodies or

The planetesimal hypothesis

"planetesimals," which in turn further coalesced into planets, their suggestion came to be called the "planetesimal hypothesis."

In 1917, the English astronomers James Hopwood Jeans (1877–1946) and Harold Jeffreys (1891–   ) worked out the planetesimal hypothesis in still greater detail and suggested that the bridge of matter pulled out between the stars would be cigar-shaped. It would be from the fattest portions of the bridge in the middle that the giant planets, Jupiter and Saturn, would form, while small planets would form beyond Saturn and within the orbit of Jupiter.

Constitution of the Sun

If the planetesimal hypothesis is accepted, it is no longer safe to assume that if 4.7 eons is the age of the Earth; it is roughly the age of the Sun as well. How long might not the Sun have existed in lonely splendor before the intruder blessed it with a family? Might not the planetary system be a relatively late addition to a Sun whose existence might be measured in tens or even hundreds of eons?

Such an extreme lifetime for the Sun became vaguely conceivable once the interconversion of mass and energy was understood. The Sun maintained its radiation at the expense of mass, but who could say, at first, how large its original mass might have been? If the Sun had originally been twice its present mass and had been losing mass constantly at its present rate, it would have existed for 1500 eons before attaining its present mass. And, of course, it could then continue to exist another 1500 eons, while radiating at its present rate, before disappearing altogether.

However, it is extremely unlikely that mass can be lost at a steady rate until it is all gone. It was the experience of physicists working with atomic nuclei that energy was usually produced at the expense of mass when one set of nuclei rearranged itself into another set of nuclei. Only a small fraction of the total mass was, under such circumstances, converted into energy. If the Sun gained its energy from nuclear reactions proceeding within itself, it could, at best, lose only a small fraction of its mass. Then, when all its matter had been rearranged into the product nucleus, nuclear reactions would stop. Little, if any, energy might be formed thereafter, even though great quantities of mass were still to be found in the Sun.

The quantity of energy available to the Sun and, therefore,

the length of time it might have existed, and might continue to exist, depended on the nature of the nuclear reactions going on within it. But how were scientists to determine the nature of these reactions? Offhand, it might seem impossible to solve such a problem unless one could first determine the nature of the substances making up the Sun's structure and the conditions under which they existed, and then try to work out the type of nuclear reactions such substances would undergo in such conditions.

Surely, this is a formidable task. To begin with how would one determine the constitution of the Sun from a distance of 93,000,000 miles? In the early nineteenth century, it might have seemed ridiculous even to dream of such a thing. Indeed, the French philosopher Auguste Comte (1798–1857) had considered what the absolute limits of human knowledge might be and, as an example of something that must forever remain unknown and unknowable, he had listed, among other items, the question of the chemical constitution of the heavenly bodies.

Yet not everything about the Sun is 93,000,000 miles away. Its radiation reaches across space and touches us. As the nineteenth century progressed, scientists learned how to squeeze more and more information out of such radiation. (The question of radial velocity, for instance, was answered by studying the radiation of stars.) Let us return, then, to the spectrum and to spectral lines.

In 1859, the German physicist Gustav Robert Kirchhoff (1824–1887) and his collaborator, the German chemist Robert Wilhelm Bunsen (1811–1899), began a careful study of the spectra produced by various vapors when heated in the virtually colorless flame produced by a "Bunsen burner" (a device, popularized by Bunsen, that mixed air and gas to promote more efficient burning and a hotter flame). The heated vapors produced an emission spectrum, bright lines against a dark background. Furthermore, the nature of these bright lines depended on the elements present in the vapor. Each element produced its own pattern of bright lines, and the same line in precisely the some position was never produced by two different elements. The emission spectrum served as a sort of fingerprint of the elements present in the glowing vapor, and thus Kirchhoff and Bunsen founded the technique of "spectroscopy."

The next year, in the course of their studies of spectra produced by various minerals, Kirchhoff and Bunsen detected lines that were not produced by any known element. They suspected the presence of new and hitherto undiscovered elements and were quickly able to verify the fact by chemical analysis. The new elements were named "cesium" and "rubidium" from Latin words meaning "sky blue" and "red" respectively, signifying the colors of the lines that led to the discovery. Cesium and rubidium were the first elements to be discovered spectroscopically, but were by no means the last.

Kirchhoff and Bunsen did more than this. They worked with light from a glowing solid (which produced white light that formed a continuous spectrum) and passed that light through cool vapor. They found that the vapor absorbed certain wavelengths of light and that the spectrum that was formed after the light had passed through the vapor was no longer completely continuous, but was crossed by dark lines which marked the position of the absorbed wavelengths. This was an "absorption spectrum," and it seemed clear at once that the Solar spectrum was an example of this. The hot body of the Sun produces white light in a continuous spectrum, and when this light passed through the Sun's atmosphere (which was hot enough but which was nevertheless cooler than the body itself), some wavelengths were absorbed; this was the reason for the dark lines in the Solar spectrum.

Kirchhoff noted that the wavelengths of light absorbed by a cool vapor were exactly the same as the wavelengths of light emitted by that same vapor when hot and glowing. Suppose, for instance, that vapor of the element sodium was heated until it glowed. The light produced would be a deep yellow; if that light were passed through a slit and then a prism, a closely spaced pair of yellow lines would appear and make up the total emission spectrum for sodium.

If, however, white light from a carbon arc were passed through relatively cool sodium vapor, the continuous spectrum ordinarily produced by the arc light would be broken by a pair of closely spaced dark lines in the yellow. The dark lines produced by absorption by the cool sodium vapor would be precisely in the position of the bright lines produced by the glowing sodium vapor. The dark lines of an absorption spectrum could serve as the identifying mark of an element as well as the bright lines in an emission spectrum could.

But then what about the Solar spectrum and the absorption lines present there? One of the most prominent lines in that spectrum (one which had been tabbed "D" by Fraunhofer) was indeed in the position of the sodium lines. Kirchhoff checked this by passing sunlight through hot sodium vapor and finding that the D line grew deeper and more prominent. Furthermore, by passing sunlight through hotly glowing sodium vapor, he could supply the sodium line and wipe out the dark D line in the Solar spectrum.

If the lines produced in the laboratory were identical with those produced in the Sun, it seemed reasonable to suppose that those in the Sun were also producd by sodium and that sodium vapor was present in the Solar atmosphere. Similarly, the dark lines H and K were shown to be produced by calcium and that must be present, therefore, in the Solar atmosphere. In 1862, the Swedish astronomer Anders Jonas Angstrom (1814–1874) showed that hydrogen was present in the Sun. Comte's dictum proved completely wrong; it was indeed possible to work out the chemical constitution of the Sun and, in fact, of any heavenly object that gave off light of its own with sufficient intensity to produce a detectable spectrum.

At first the Solar spectrum was used only as a means of determining which elements were present in the Sun and which were not. But the question "How much?" arose. Spectral lines deepened and broadened with increasing concentration of a particular element in the glowing or absorbing vapor. It began to be possible to determine not only whether an element were present but in what quantities it might be present.

Finally, in 1929 the American astronomer Henry Norris Russell (1877–1957) carefully studied Solar spectra and was able to show that the Sun was astonishingly rich in hydrogen. He decided that the hydrogen content of the Sun made up fully three-fifths of its volume. This was totally unexpected since hydrogen, while not exactly rare, makes up but a small portion of the Earth's crust, only 0.14 percent, in fact.

Yet later work showed that Russell was overconservative. The recent estimates of the American astronomer Donald Howard Menzel (1901–1976) show that 81.76 percent of the Sun's volume is hydrogen and 18.17 percent is helium. That leaves only 0.07 percent for all other atoms.

It seems safe to say, then, that the Sun is essentially a glowing mixture of hydrogen and helium, in a ratio of about 4 to 1 in terms of volume. (The element helium was another one discovered spectroscopically — and in the Sun, rather than on Earth. The English astronomer Joseph Norman Lockyer (1836–1920) suggested that certain unknown lines in the Solar spectrum might be produced by an as-yet-undiscovered element which he named after Helios, the Greek god of the Sun. It was not until 1895 that helium was located on the Earth by the Scottish chemist William Ramsay [1852–1916].)

Surface Temperature of the Sun

Knowledge concerning the constitution of the Sun drastically reduced the number of nuclear reactions that might serve as conceivable sources for the Sun's vast production of energy. It was simply out of the question to suppose that the main source could arise out of nuclear reactions in which any substance other than hydrogen, or possibly helium, might serve as the fuel. No other substance was present in large enough quantities.

Suppose, then, we consider the atomic nuclei of hydrogen and helium. The nucleus of the most common type of hydrogen atom consists of a single particle, a proton, and this type of hydrogen atom is therefore called "hydrogen-1." The nucleus of the most common type of helium atom is made up of four particles, two protons and two neutrons, and the atom is therefore referred to as "helium-4."

It is conceivable that four hydrogen nuclei might fuse ("hydrogen fusion") to form a single helium nucleus, a process which we might represent as: $4H^1 \longrightarrow He^4$. Without going into the details of how this might come to be, either directly or through a long series of reactions involving other atoms, let us ask only whether such hydrogen fusion would suffice to supply the Sun with the necessary energy.

The mass of a hydrogen nucleus has been determined with great accuracy in "atomic mass units." In these units, the mass of the hydrogen nucleus is 1.00797; four such nuclei would have a mass of 4.03188. The mass of the helium nucleus, however, is only 4.0026. If 4.03188 atomic mass units of hydrogen are somehow fused into 4.0026 atomic mass units of helium, then 0.0293 atomic mass units (0.73 percent of the whole) must be converted into energy.

The Sun's loss of 4,600,000 tons of mass each second (see page 108) could, in that case, represent a mass loss resulting from the conversion of hydrogen into helium. Hydrogen would be the Sun's nuclear fuel, helium its nuclear "ash." Since the mass loss as the result of the conversion of hydrogen into helium is 0.73 percent of the mass of the fusing hydrogen, the loss of

4,600,000 tons of mass each second means that each second 630,000,000 tons of hydrogen are being converted into helium.

This fact makes it possible to guess at what the age of the Sun might be. The total mass of the Sun can be calculated from the strength of its gravitational pull on the Earth across a distance of 93,000,000 miles; this turns out to be 2,200,000,000,000,-000,000,000,000,000 tons. Each second 630,000,000 tons of hydrogen are being consumed; and if we assume that the Sun was pure hydrogen to start with, that this hydrogen has been fusing at a constant rate since the beginning, and the material of the Sun is always well mixed, we can calculate how many seconds it would take before the quantity of hydrogen sinks from 100 percent to 81.76 percent. It turns out that this would take 20,000,000,000 years — that is, 20 eons. Furthermore, it would take an additional 90 eons before the total hydrogen fuel that remains is consumed.

It is, of course, unsafe to assume that the rate of fusion will continue unchanged to the end of the fuel supply or that it was always the same as it is now. Certainly, the presence of different quantities of helium ash might be expected to influence the rate and even, perhaps, the nature of the reaction. Nevertheless, as the 1930's dawned, it seemed at least possible that a plausible scheme might be drawn up that would account for a Sun with a total lifetime of at least 100 eons. The Solar system clearly had a vast stretch of past history, and an even vaster stretch still lay ahead of it.

Merely conjecturing that the Sun's energy supply was based on the fusion of hydrogen to helium was insufficient. It was also necessary to show that conditions on the Sun were such that hydrogen would indeed fuse. There are vast supplies of hydrogen here in the Earth's ocean, for instance, and yet it does not undergo fusion. If it did, the Earth would explode and vaporize, changing into a very small and very temporary star. If the fusion proceeded in a controlled fashion, however, man might be able to supply his energy needs for millions of years to come. However, conditions on Earth are not such that spontaneous hydrogen fusion is possible, nor can scientists — so far — create the necessary conditions for controlled fusion. The best they have been able to do, is to force some hydrogen to undergo uncontrolled fusion, thus producing the "hydrogen bomb" of the 1950's.

What, then, about conditions of the Sun?

All we can see of the Sun is its surface, and that surface, it is obvious, is hot. But just how hot? Again we turn to its radiation.

Even an object that is quite cool, such as the human body, is constantly radiating energy. The body's warmth can be felt from a small distance. But this radiation is very long-wave in nature and is in the far infrared and therefore completely undetectable by the eye.

If an object, such as a flatiron, is slowly heated, the radiation it gives off becomes more copious and begins to spread into the range of shorter and shorter wavelengths. Photographic film capable of responding to infrared radiation can take a picture of a hot flatiron in a dark room by the flatiron's own "light" even though that light is still too long-wave to be detectable by eye. Eventually, if the flatiron were heated still further, some of the radiation would appear in wavelengths so short as to be detectable to the eye. They would be made up of the longest visible wavelengths to begin with, and so the object would glow a deep red. Further heating would add more and more radiation of shorter and shorter wavelength, and the color would change correspondingly.

In 1893, the German physicist Wilhelm Wien (1864–1928) studied the matter in detail. At any given temperature there was some radiation peak, some wavelength at which more radiation was emitted than at any other. Wien found that as the temperature increased, the position of this peak shifted in the short-wave direction according to a simple mathematical rule. If, then, the spectrum of any glowing object is studied and if the radiation peak in that spectrum can be determined, the temperature of the object can be deduced. The nature of the spectral lines also changes with temperature, and they, too, will serve as evidence in this respect.

Working with the Solar spectrum, it could then be shown that the temperature of the Sun's surface is 6000° C. or 10,000° F.). The surface temperatures of other stars can also be determined in similar fashion, and some turn out to be hotter than the Sun. The surface temperature of Sirius is 11,000° C., for instance, and that of Alpha Crucis (the brightest star of the Southern Cross) is 21,000° C.

In earthly terms, the Sun's surface is very hot. It is hot enough to melt and vaporize all known substances. Nevertheless, it is not hot enough — it is very far from hot enough — to force hydrogen to undergo fusion into helium. We can say with perfect safety that no such fusion takes place anywhere on the Sun's surface, yet it must take place somewhere if the Sun's energy output is to be explained. That leaves us, then, with the question of what might be going on in the Sun's interior.

Internal Temperature of the Sun

Determining the properties of the Sun's surface was, indeed, a formidable achievement and one that, on the face of it, might have seemed impossible. How much more difficult might one suspect an attempted study of the Sun's interior to be.

Nevertheless, some conclusions concerning the Sun's interior are easily arrived at. Thus, the Sun's surface is continually losing heat to outer space, and at a great rate, too; yet the temperature of the surface remains constant. Clearly, the surface must be gaining heat from the interior as fast as it loses it to space, and it follows that the interior must be hotter than the surface.

Since the Sun's surface is already hot enough to vaporize all known substances and since the interior of the Sun is hotter still, it seems reasonable to suppose that the Sun is gaseous in nature — just a globe of superhot gas. If this is so, then astronomers are fortunate, for the properties of gases are easier to understand and work with than are the properties of liquids and solids.

During the 1920's, the problem of the structure of the Sun's interior was tackled by the English astronomer Arthur Stanley Eddington (1882–1944), working on the assumption that stars were gaseous bodies.

He reasoned that if the Sun were affected only by its own gravitational field, it would collapse if it were merely a globe of gas. The fact that it did not collapse meant some force was countering that gravity, a force exerted from within outward. Such an outward-directed force might be that produced by the expansive tendencies of gases at high temperature.

Taking into account the mass of the Sun and the strength of its gravitational field, Eddington in 1926 calculated the temperatures required to balance the gravitational force at the different depths beneath the Sun's surface and ended up with amazing figures. The temperature at the Sun's center reached the colossal figure of 15,000,000°C. (Some recent calculations, indeed, would put this figure as high as 21,000,000°C.)

Despite the astonishing nature of this finding, it was accepted by astronomers generally. For one thing, such temperatures were necessary if hydrogen fusion were to be possible. While the Sun's surface was far too cool for any hydrogen fusion reaction to proceed, the Sun's interior, by Eddington's reckoning, was emphatically hot enough for the purpose.

Then, too, Eddington's reasoning helped explain certain other phenomena. The Sun was in a state of delicate balance between gravitational force pulling inward and the temperature effect pushing outward. What if some other star were not?

Suppose, instead, that a particular star were not quite hot enough to oppose gravitational compression. Such a star would collapse inward, and this collapse would indeed (as Helmholtz had long ago argued) convert gravitation energy into heat. The interior temperatures would begin to rise and expansive forces would intensify and eventually reach the point where they would just balance gravitational pressure. Inertia, however, would keep the star contracting for a while past that balance point — but more and more slowly. By the time the contraction was brought to a complete halt, the temperature would be considerably higher than was needed merely to counter the gravitational pressure, and the star would begin to balloon outward. As it expanded, the temperature would drop and quickly fall to a point of balance again. But here, too, inertia would keep the star ballooning past that point until it slowed and began to contract again. This cycle would continue over and over indefinitely.

Such a star would pulsate about some equilibrium point, much like a swinging pendulum or bobbing spring. Naturally the brightness of such stars would change regularly with the pulsations and in such a way (taking into account both size and temperature) that the light variation would just match the behavior of Cepheid variables.

With the temperature and pressure of the Sun's interior decided on and accepted, it remained to work out the manner in which hydrogen would fuse to helium under such circumstances at just the rate that would account for the quantity of radiation being emitted by the Sun. In 1939, the German-American physicist Hans Albrecht Bethe (1906–   ) was able to work out a series of nuclear reactions that just filled the bill. The rate at which they would proceed under the conditions of the Sun's interior (as calculated from data, both theoretical and observed, gained in the physics laboratories of the Earth) fit the conditions very well.

The question of the source of the Sun's energy, posed by Helmholtz in the 1840's, was thus finally settled by Bethe nearly a hundred years later. And with that, a potential 100-eon lifetime for the Sun was also established.

The gathering evidence in favor of a superhot interior of the Sun had, however, an unexpected side effect. It ruined the planetesimal theory of the origin of the Solar system.

It was all very well to suppose that matter could be pulled out of the Sun and that this would then condense and form planets — as long as the Solar material was assumed to be at a temperature of merely thousands of degrees. Temperatures in the millions of degrees were another matter entirely.

In 1939, the American astronomer Lyman Spitzer, Jr. (1914–   ) advanced what seemed valid arguments for concluding that such superhot matter would never condense to form planets but would spread out rapidly to form a gaseous nebula about the Sun, and would stay so.

Astronomers therefore had to return to the task of attempting to visualize the formation of the planets out of relatively cool matter. They had, once again, to think of a contracting nebula

Weizsäcker's hypothesis

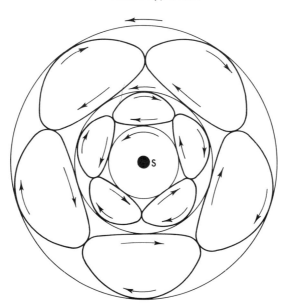

in the old Laplacian fashion. However, much had been learned in the twentieth century about how such a nebula might be expected to behave and about the electrical and magnetic forces to which it would be exposed, in addition to the gravitational ones.

In 1943, the German astronomer Carl Friedrich von Weizsäcker (1912–   ) suggested that the nebula out of which the Solar system was formed did not revolve as a unit. Instead turbulent patterns would be set up in the outer reaches, smaller whirlpools within the overall larger ones. Where adjacent whirlpools met each other, there would be collisions of particles and coalescence into larger and larger particles and, eventually, the gathering of planets. In this way, Weizsäcker attempted to account for all that Laplace tried to account for and, in addition, to explain the spacing of the planets, the distribution of angular momentum, and so on.

Weizsäcker's theory was greeted enthusiastically, but there was considerable controversy over many of the details. The controversy continues, and a number of astronomers have advanced their own version, no one of which has as yet met general acceptance. The Swedish astronomer, Hannes Alfven (1908–   ) and the English astronomer, Fred Hoyle (1915–   ) have, however, suggested mechanisms involving the Sun's magnetic field for transferring angular momentum to the planets. These mechanisms carry considerable weight of plausibility and have proved impressively popular.

All theories agree, however, in postulating the formation of the entire Solar system, both Sun and planets, by a single process. In other words, if Earth, in its present shape, is 4.7 eons old, then we may conclude that the entire Solar system (including the Sun), in the form it now exists, is 4.7 eons old.

9

Types of Stars

Constitution of the Solar System

The process of hydrogen fusion to helium, while serving as the main answer to the question of the Sun's energy production, does not remove all problems. For one thing, the Sun turns out to be unexpectedly poor in hydrogen and rich in helium. If it has been existing only 5 eons or so, it ought to have expended less hydrogen than it has and formed less helium.

It could be, one might suppose, that the Sun was hotter in the past and spent its fuel with a more liberal hand. That seems a natural thought, in fact, since one might expect the Sun to behave like a wood fire, its flame sinking as its fuel is consumed and "burning" progressively more slowly. In that case, its past is shorter than we might think, but its future is correspondingly longer. The chief trouble with this suggestion is that from what geologists can guess or deduce concerning the Earth's past history, there has been no significant change in the Sun's radiation output in the last few eons.

A second possibility is that the Sun was expending hydrogen before the formation of the Solar system, while it was still an extended, turbulent nebula:

This is also unlikely. The nebula might, conceivably, have existed for an indefinite number of eons before the formation of the Solar system in its present form, but it would not be losing energy at the expense of nuclear reactions while it was a nebula. In an extended nebula, the gravitational field is so diffuse that it brings about very little temperature increase toward the center, not enough to reach the ignition point that would start the hydrogen fusion going. Such a nebula would slowly contract and only the gravitational energy of particles falling inward would be available for energy production — after the old fashion postulated by Helmholtz.

As the nebula contracted, the gravitational field would grow more intense; the total energy involved would remain the same but it would be concentrated into a smaller and smaller volume. As the pressures at the center of the contracting nebula increased, so would the temperature, until finally, the ignition point would be reached. At this point, the collapsing nebula would fire up and become a star. It is only then that nuclear reactions would take place, and only at the Sun's center, not in the outskirts where the planets were taking shape.

We are left then, not only with the problem of the too-abundant helium, but with a quantity of elements still more complicated than helium that also exist in the Sun and in the planets. Where did the other elements come from?

Let us consider the various elements for a moment. Hydrogen, with its atomic nucleus made up of a single particle, and helium, with an atomic nucleus made up of four particles, are the two simplest elements. The remaining elements are all more complicated. The most common elements (next to hydrogen and helium) are carbon, nitrogen, oxygen, and neon, and their atomic nuclei are made up of twelve, fourteen, sixteen, and twenty particles, respectively.

One might suppose, of course, that while hydrogen fused mainly to helium, there might be certain side reactions in which helium atoms would fuse further to carbon, say, or to oxygen. These further fusion reactions must be rare indeed since they have sufficed to produce only very small quantities of the more complex atoms even over the nearly 5-eon history of the Sun. Oxygen, for instance, makes up only 0.03 percent of the volume of the Sun.

Then, too, if this were so, the elements beyond helium would exist only within the Sun as the result of their formation from nuclear fusion reactions. How could so much of the more complex atoms find their way into the planets which were formed out of matter at the outskirts of the nebula?

Earth, for instance, is formed almost exclusively out of elements more complicated than hydrogen and helium. This is not quite as surprising as it might sound at first; there are some excuses that can be made here.

Solid substances cohere by interatomic attractions and do not depend on gravitational force to hold together. Gases and vapors, however, have only very small interatomic attractions, and it takes gravitational force to hold them to the body of a planet. The motions of the atoms or atom-groupings (the latter are called "molecules") in gases and vapors tend to counteract gravitational pull. If the atoms or molecules move quickly enough, they leave the vicinity of a planet despite its gravitational pull. The smaller the planet and the weaker its gravity, the more easily do atoms and molecules drift away. Then, too, the lighter the atom or molecule, the more rapidly it tends to move, and the more likely it is to drift away.

Hydrogen atoms are the lightest of all. They tend to double up to form hydrogen molecules. Although the hydrogen molecule has twice the mass of the single hydrogen atom, it is still lighter than any other atom.

Helium occurs as single atoms. The helium atom is twice as massive as the hydrogen molecule (and four times as massive as the hydrogen atom), but it is lighter than any atom or molecule other than those of hydrogen.

The Earth's gravitational field is not strong enough to hold either hydrogen or helium. In the case of hydrogen, there are alleviating factors. Two hydrogen atoms can join with a single oxygen atom to form a water molecule (eight times as massive as a hydrogen molecule), and can join with other atoms to form molecules of solids. As a result, the Earth in the course of its formation, did manage to retain some hydrogen in combination with other elements, but its gravitational field was never large enough to hold hydrogen in its gaseous form. As a result, most of the hydrogen that must have been in the vicinity of the Earth as it formed was never captured, and that is one reason why the Earth is as small as it is. As for helium, it forms no compounds at all, so not even a respectable remnant of helium was captured. Earth today possesses only tiny traces of helium.

And yet quantities of other elements (chiefly oxygen, silicon, and iron) remained to form a body the size of the Earth, and other bodies the size of Mars, Venus, Mercury, and the Moon.

A planet like Jupiter, much farther from the Sun than we are, may always have been at a much lower temperature. The lower the temperature, the more slowly do atoms move and the easier they are to hold. The matter gathering to form Jupiter could hold hydrogen more easily than the matter gathering to form Earth. As more hydrogen was collected, the mass of Jupiter rose and so did the intensity of its gravitational field. It became still easier to collect further hydrogen, which, in turn, heightened the gravitational field still more. It was this snowball effect that allowed Jupiter to grow so large and, as spectroscopic and other evidence tells us, to be so rich in hydrogen (as are, indeed, the other cold, outer worlds).

Nevertheless, even Jupiter is not all hydrogen. The atmosphere has a large admixture of helium, and there is also evidence of the presence of compounds containing carbon and nitrogen.

So there were surprising quantities of helium and more complex elements throughout the nebula out of which the planets were formed. The alternatives are as follows:

1. The presence of the heavy elements is impossible except within the Solar interior, and therefore the planets must have originated from matter within the Sun. This would argue against any nebular theory of origin and force astronomers back to some version of the planetesimal theory.

2. The presence of the heavy elements within the extended nebula is possible, and they must have been formed by some method other than nuclear reactions within the Solar interior.

Most astronomers are reluctant to accept the first alternative if they can somehow work out the second in acceptable detail. To see where the heavy elements might come from, if the Sun is left out of consideration, let us turn our eyes beyond the Solar system and look to the stars once more.

Spectral Classes

The earliest differences noted among the stars were in connection with their position and brightness. In a few cases, there were color differences, too. Antares was red, Capella yellow, Sirius white, and Vega bluish-white. Such colors were noticeable to the naked eye only among a small handful of very bright stars.

The first half of the nineteenth century added an important difference in distance. Some stars were relatively close by (only a hundred trillion miles or so), while others were enormously farther off. This meant that one could calculate the actual brightness or luminosity of those stars whose distance was known, and important differences in luminosity were noted.

Once spectroscopy came into use in the second half of the nineteenth century, it was natural to wonder whether the different stars might produce different types of spectra. The Italian astronomer Pietro Angelo Secchi (1818–1878) studied the spectra that were available to him and in 1867 suggested that they might be divided into four classes. The Solar spectrum fell into his second class, one that was characterized by the presence of numerous absorption lines of metals such as iron.

Later astronomers confirmed the existence of these "spectral classes" and refined them, introducing more delicate divisions. By 1900, the American astronomer Edward Charles Pickering (1846–1919) was characterizing the classes by letters of the alphabet. The Sun was in spectral class G, for instance. It was eventually possible to classify the spectra within each class by numbers from 0 to 9, and the Solar spectrum was classified as G2.

The spectral classes were not widely different. Rather, one faded into another to form a sort of continuum. It seemed probable from that, that whatever difference in properties gave rise to the different spectral classes, it was a continuous one, and that the property that changed did so smoothly.

The question is: What property change accounts for the difference in spectra?

Kirchhoff and Bunsen had shown that each element produced its own characteristic spectrum. Therefore, if spectra of two stars differed, should that not be because the two stars consisted of different sets of elements? This was not an attractive thought. While it was possible that one star might possess elements another did not, that did not fit in with the growing conception of all objects in the Universe being made up of the same, (rather limited in number) elements.

Was it possible, instead, for spectra to undergo alterations without any essential change in elemental makeup of the body yielding the spectrum?

One method of producing this effect was to alter the temperature. With changing temperature, the electrons surrounding the atomic nuclei shift from one "energy state" to another. As the temperature rises, an electron will shift from a lower energy state to a higher one and absorb a certain wavelength of light as it does so. It may later shift back from the state of higher energy to one of lower, and light of that same wavelength is then emitted. Because electrons can shift from state to state in a number of ways, a particular type of atom will emit or absorb a number of different wavelengths, forming a spectral pattern of bright lines or dark ones, the patterns being the same in either case.

Every element has atoms containing a distinctive number and arrangement of electrons. The electrons in each different kind of atom have their own distinctive spectral pattern therefore, not shared by any other kind of atom with a different number and arrangement of electrons. It is for this reason that spectral lines, either bright or dark, can be used to identify elements.

The hydrogen atom contains but a single electron, and its spectral pattern is comparatively simple since there is only so much a single electron can do. As atoms grow more and more complicated, with more and more electrons in the atom, the pattern grows more and more complicated, too. Sometimes the spectral pattern does not appear to be as complicated as one might expect because most of the lines are outside the visible range. The iron atom, however, containing twenty-six electrons, produces thousands of lines in the visible range. It is iron that is responsible for the major portion of the complexity of the Solar spectrum in the visible range.

If one continues to heat a substance and forces the electrons of the atoms making it up to enter states of higher and higher energy, there eventually comes a point at which some of the electrons become sufficiently energetic to break the hold of the central nucleus and leave the atom. As temperatures increase, one electron after another will leave the atom.

An atom with fewer electrons than its normal complement (or, for that matter, more than its normal complement) is called an "ion." The loss of electrons is therefore referred to as an "ionization."

An ionized atom will produce a different spectral pattern from that produced by a normal atom. With one or more electrons gone, those that remain shift among energy levels in a somewhat different manner. Furthermore, an atom with one electron missing will not give the same pattern as will that same atom with two electrons missing, or three.

Different kinds of atoms hold on to their electrons with

different strengths. A temperature that suffices to ionize a sodium atom, let us say, is completely insufficient to ionize an oxygen atom. Again, it always takes a higher temperature to drive off a second electron from an atom than it did to drive off the first one, and a still higher temperature is required to drive off the third electron, and so on.

In short, differences in spectra may reflect not differences in elements, but differences in the ionization state of the elements. This, in turn, will reflect a difference in temperature.

Before this was understood, strange lines were sometimes located in spectra and attributed to unknown elements. This worked out well in the case of helium (see page 117), but in no other instance. Thus, the element "coronium" was reported in the Sun's outer atmosphere — a region called the "corona" and visible only during a total eclipse. Similarly, an element called "nebulium" was reported as present in certain nebulae.

In 1927, however, the American astronomer Ira Sprague Bowen (1898–   ) showed that the lines attributed to nebulium were actually produced by a mixture of the long-known elements, oxygen and nitrogen, where the atoms of each had lost a couple of electrons under conditions that required very low densities. Then, in 1941, the Swedish astronomer Bengt Edlen (1906–   ) showed that coronium was actually a mixture of iron and nickel atoms that had lost about a dozen electrons apiece.

Once spectra came to be interpreted in the light of ionization, it became possible to tell from the pattern of the lines alone, the temperature of the surface of the star being studied. The difference between the spectral classes was taken to be that between stars at different temperatures and to only a very minor extent that of stars made up of different elements. Indeed, the constitution of the vast majority of stars is remarkably uniform. Like the Sun, most stars are made up largely of hydrogen and helium.

If the spectral classes are arranged in order of decreasing temperature, the letter designations read: O, B, A, F, G, K, and M. There are four additional classes, rather specialized ones, which are labelled R, N, S, and W (the first three including cool stars, the last hot ones).

Giant Stars and Dwarf Stars

Once one obtains two classes of knowledge concerning a number of different stars, such as their luminosity and their surface temperature, the next logical step is to put the two together. For instance, on the basis of experience with glowing objects on Earth, we would expect that the cooler a star was, the less radiation it would emit, and the dimmer and redder it would be. This, however, proved not always to be so.

For instance, if the temperature-interpretation of spectral classes is accepted, then the coolest of the ordinary stars are those of spectral class M. From the spectral lines they exhibited and from the location of their radiation peak, it was estimated that a typical surface temperature for this spectral class was 2500° C. as compared with our Sun's 6000° C. The M-class stars were, indeed, uniformly reddish in color, but they were not — despite expectations — uniformly dim. Many of them were, indeed, dim stars even though a few, like Barnard's star, were quite close to us. Others, however, like Betelgeuse in Orion, or Antares in Scorpio, were red in color but very bright in appearance just the same. Nor was this because they were particularly close. They were not only bright in appearance, they were very luminous in actual fact. Antares, for instance, emits some ten thousand times as much radiation as the Sun.

As early as 1905, E. Hertzsprung had speculated on this matter and had decided that the only way in which a cool star could be bright was to be enormously large. Its coolness would mean that its surface gave little light per square mile as compared with the Sun but, on the other hand, there would be vastly more square miles on a star like Betelgeuse than on the Sun. The number of square miles would more than make up for the comparative dimness of each one individually. For this reason, stars like Betelgeuse and Antares came to be called "red giants," while stars like Barnard's star were "red dwarfs." It was particularly interesting that there seemed to be no "in-between" red stars that were neither giants nor dwarfs. This lack of red stars of intermediate size is called the "Hertzsprung gap."

This suggestion by Hertzsprung, based on theoretical reasoning, was shown to be correct by actual observation. The German-American physicist Albert Abraham Michelson (1852–1931) invented a device called the "interferometer" in 1881. By detecting slight variations in the manner in which light waves interfered with each other, this device was capable of making astonishingly refined measurements. It made it possible, for instance, to learn things about stars that no telescope could reveal.

Even the nearest stars are so distant that the best modern

telescope does not make them seem large enough to be more than a point of light. Nevertheless, the rays of light that reach the telescope from the star do not all come from the same point on the star. One ray may come from the star's eastern edge, another from the star's western edge. The two reach the telescope at a small angle to each other, an angle too small to measure by ordinary methods but sometimes large enough to allow the rays to collide, so to speak, and interfere with each other. Michelson's instrument made it possible to measure the extent of this interference and to determine the angle, if it were not too tiny. If the angle is known, and the distance of the star as well, then the actual diameter can quickly be determined.

The results were astonishing. The diameter of Betelgeuse was measured in this fashion in 1920 and found to be about 300,-000,000 miles. If this is compared with the Sun's diameter of 865,000 miles, it can be seen that Betelgeuse is nearly 350 times as wide as the Sun. It therefore was 350 × 350 or about 120,000 times the surface area of the Sun. No wonder it is much more luminous than the Sun even though it is dimmer by the square mile. As for its volume, it is approximately 40,000,000 times that of the Sun. If Betelgeuse were put in place of the Sun, its mighty diameter would cause it to fill out all space to beyond the orbit of Mars. It is a red giant indeed.

Antares is somewhat smaller than Betelgeuse, but the latter star is by no means a record-holder. One star of this sort, Epsilon Aurigae, is so cool that, despite a monstrous size, it is completely invisible to us. It radiates almost entirely in the infrared. We know of its presence only because it has a bright companion which is periodically eclipsed by it. From the duration of the eclipse and the distance of the system, it was suggested, in 1937, that the dark star is an "infrared giant" that is 2,300,000,000 miles in diameter. If it were inserted into the Solar system in place of the Sun, it would fill space out to the orbit of Uranus.

Nor are the infrared giants as rare as they seemed at first. The trouble is that if a star is so cool that it radiates mostly in the infrared, it becomes very hard to detect. In the first place, the Earth's atmosphere is not very transparent in the infrared region, and secondly everything on Earth itself is warm enough to emit considerable infrared radiation of its own, so that infrared radiation from the sky tends to be lost in the glare, so to speak.

In 1965, however, astronomers at Mount Wilson Observatory began to make use of special techniques, including a telescope with a 62-inch plastic mirror, with which to scan the sky for spots rich in infrared radiation, spots that would indicate the presence of infrared giants. Within a couple of years, they found thousands of such objects, mostly concentrated in the plane of the Milky Way. Some of them, at least, may be even more voluminous than Epsilon Aurigae. They are bright indeed in the infrared region but extremely dim in the visible range, and few are visible in even the largest telescopes. Two stars which have been detected are judged, from their colors, to have temperatures of 1200° K and 800° K, respectively — the second just *barely* red hot.

Stars of other colors do not show quite the gap in size among

Sizes of the Stars

Star	Spectro-scopic Type	Measured Diameter* Sun = 1	Calculated Diameter Sun = 1	
ε Aurigae B	K5	2000		⎫
VV Cephei A	M2	1200		
o Ceti	M6e	*400*		
α Orionis	M2	*300 — 400†*	400	Supergiants
ε Aurigae A	K5	300		
α Scorpii	M1	*300*	320	
ε Aurigae A	F5	200		
β Pegasi	M5	*110*	130	⎭
α Tauri	K5	*36*	57	⎫
V 380 Cygni A	B2	*29*		
α Boötis	K0	*23*	26	Giants
α Aurigae	(G0)		16	⎭
Y Cygni A, B	O9	*5.9*		
β Persei A	B8	3.1		
β Aurigae A	A0	2.8		
Procyon A	F5		1.7	⎫
α Centauri E	G4		1.2	
Sun	G0	1.0	1.0	
W Ursae Majoris A	F8	0.9		Main Sequence Dwarfs
70 Ophiuchi A	K1		0.9	
70 Ophiuchi B	K5		0.7	
Krüger 60 A	M4		0.5	⎭
Sirius B	A5		0.02	⎫
40 Eridani B	A		0.02	White
van Maanen 2	G		0.006	Dwarfs
Wolf 457	(a)		0.003	⎭

* Values in italics are derived from Pease's interferometric measurements modified to allow for the best recent parallax measurements. Other figures are derived from the light curves of eclipsing biniaries.
† Variable diameter.

themselves that the cool, red stars do. Still, there are large "yellow giants" (not as large or cool as the red ones) and small "yellow dwarfs" (not as small or cool as the red ones). Capella could be described as a yellow giant, our Sun as a yellow dwarf.

The H-R Diagram

While Hertzsprung was discovering the red giants, H. N. Russell was doing similar work. He prepared a graph in 1913 (independently of Hertzsprung, who had done the same thing some years before) in which the horizontal axis was made up of the spectral classes in order of descending temperature, starting with spectral class 0 on the left and ending with spectral class M on the right. The vertical axis represented the luminosity or absolute magnitude (see page 41). Each star has some absolute magnitude and some spectral class and could be represented by a dot, therefore, in a particular position on the graph. Such a graph is the "Hertzsprung-Russell diagram" or, more commonly, the "H-R diagram."

In general, the hotter the star, the brighter it is. On the graph then, the farther left a star was in spectral class (and there-

H-R diagram

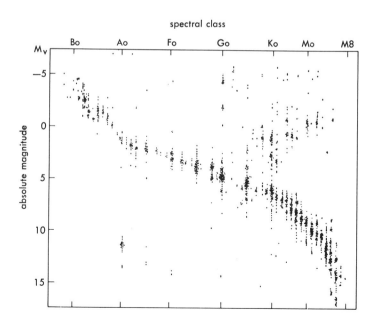

spectral class

fore higher in temperature) the higher up it was in absolute magnitude. As a result, most of the stars plotted by Russell fell into a diagonal line stretching from the upper left to the lower right. This forms the "main sequence." It is estimated now that over 99 percent of the stars we can observe fall somewhere on this main sequence.

The most conspicuous exceptions to this rule are, of course, the red giants. They are of spectral class M and are therefore to the right of the diagram. They are also of high luminosity, however, and are clustered in the upper right of the H-R diagram, well out of contact with the main sequence.

When the H-R diagram was first prepared, notions about nuclear reactions within stars were still quite vague, and most astronomers were still thinking of stars in terms not too far removed from the views of Laplace and Helmholtz. The feeling generally was that stars systematically and steadily contracted in the course of their lifetime. From this standpoint it seemed that the H-R diagram offered a clear and dramatic picture of "stellar evolution"; the manner, that is, in which stars came into being, passed through various stages, and finally ceased to radiate.

Russell's suggestion as to the significance of the H-R diagram can be summarized as follows:

A star begins as a vastly voluminous conglomeration of cool gas which slowly contracts. It warms as it contracts and at an early stage radiates chiefly in the infrared so that it is an infrared giant like Epsilon Aurigae. It contracts further and grows hot enough to glow a bright red, like Betelgeuse and Antares. It continues to shrink and heat up, becoming a yellow giant, smaller and hotter than the red giant, and then a "blue-white star," still smaller and still hotter.

A blue-white star of spectral class O is only moderately larger than the Sun, but much hotter, with surface temperatures up to 30,000° C., five times that of the Sun. At this point the radiation peak is in the blue-violet region of the visible spectrum, and beyond that in the ultraviolet — hence the color of the star.

In traveling from the initial cool nebula to the blue-white stage, the star has been moving leftward across the top of the H-R diagram. At the blue-white stage it reaches the upper left-hand edge of the main sequence.

Now the star is pictured as continuing to contract under the influence of gravity, but for some reason it no longer grows hotter. One early suggestion was that by the time the blue-white stage was reached, the material at the center of the star was compressed so tightly that it no longer acted as a gas. With further contraction, more and more of the center would be compressed beyond the gaseous stage and that might, for some reason, progressively cut down the production of heat.

Therefore, the blue-white star shrinks and cools, growing rapidly dimmer for both reasons. It becomes a yellow dwarf like our Sun, then a red dwarf, like Barnard's star, and finally blinks out altogether and becomes a "black dwarf," a burnt-out cinder of a star.

In shrinking from a blue-white star to the final stage of black dwarf, the star slides down the main sequence from the upper left to the lower right. We might refer to this as the "slide-theory" of stellar evolution.

This scheme was a most attractive one, and there seemed to be a great many plausible things about it. First, the picture of steady shrinkage with heating at first and then cooling seemed in line with what one would expect to happen "naturally." As gas was compressed in the laboratory, it grew hotter; as hot objects were allowed to stand, they grew cooler.

Then, too, if a star is a red giant at some early stage in its career and a red dwarf at some late stage, one should expect to find red dwarfs not very much different in mass, on the average, from red giants. In other words, a red giant is not enormous because it contains a great deal of matter, but only because what matter it does contain is spread out over vast space. This proved to be so. The red giants are by no means as enormously massive as one might expect from their sheer size but were, instead, enormously thin. The matter of a star like Epsilon Aurigae would (through most of its volume) be considered merely a vacuum if a portion of it could be transported, unchanged, into the laboratory.

In fact, stellar masses are surprisingly uniform. Whereas stars vary widely in volume, density, temperature, and other properties, they do not vary much in mass. Most stars have masses that range from 1/5 to 5 times that of the Sun.[1]

[1] The masses of individual stars are determined most easily when those stars are part of a binary-star system. If the distance of the system from ourselves is known, then astronomers can work out the actual distance of the individual stars of the system from each other, and determine, also, their orbital speed. From this, by use of Newton's law of universal gravitation, the mass of the stars can be determined. Fortunately, there are so many binary stars that the data on stellar masses are large enough to allow general conclusions.

10

Stellar Evolution

The Mass-Luminosity Relation

This interesting picture of stellar evolution as a slide down the main sequence did not, however, survive long. It was dead within a decade.

By the slide theory, the Sun ought to be in a late stage of its evolution, long past its greatest, hottest days. It had, according to the slide-theory, already cooled from a blue-white star to a yellow dwarf, with red dwarfhood and final extinction perhaps not too far ahead (on a cosmic time-scale). Yet as it came to be understood that hydrogen was the most likely stellar fuel and that hydrogen was present in the Sun in overwhelming quantities, it became clear that the Sun must have a long lifetime ahead of it and must be a relatively young star in terms of that lifetime. Any evolutionary scheme that makes the Sun into an old star cannot be right.

Then, too, the question of stellar masses grew increasingly important. It is true that there are no *enormous* differences in mass between large, bright stars and small, dim ones, but it is also true that the luminous stars are somewhat more massive than the dim stars. This moderate difference, which is quite a consistent one, must be explained.

The slide-theory of stellar evolution could manage it. One

139

could argue that a large star had a greater fuel supply and therefore lasted longer. A small star would run through its fuel quickly and reach the red dwarf stage, while a large star might still be at the blue-white stage.

This explanation broke down, however, as a result of Eddington's researches on the structure of stellar interiors (see page 121).

Eddington argued in 1924 that temperature from within had to balance the compressing effect of gravitational forces. The more massive a star, the greater the gravitational compression and therefore the higher the countering temperature, the more luminous it must be. This "mass-luminosity relationship"[1] sets an upper and a lower limit to the mass of a star. Too little mass and the star cannot radiate visible light at all; too much mass and the radiation pressure from within would be sufficient to disrupt it. The greatest mass a star can have and remain intact is about 65 times that of the Sun. The most massive known stars are only half that size.

This means that one cannot picture all stars as starting at the extreme upper right hand of the H-R diagram and traveling leftward (compressing and growing hotter as they do so) to the upper left-hand portion of the main sequence at Class O. This could only hold true for particularly massive stars. Less massive stars would not need to attain Class O temperatures to support their structures against gravitational compression. They would develop lower temperatures and would reach the main sequence at levels below Class O. A star no more massive than our Sun would reach the main sequence at its present position.

Red dwarfs would be of still smaller mass to begin with and would reach positions on the main sequence even lower than the Sun. Particularly small bodies, less than 1/100 the mass of the Sun, may be incapable of raising the central temperature to the point of igniting hydrogen fusion at all. Such bodies would condense to a cold, solid structure and become a black dwarf, not as the dead cinder of a once-glowing star, but as a body that

[1] The mass-luminosity relation explained the connection between luminosity and period among the Cepheid variables. The more massive a Cepheid, the more luminous it is and also the more ponderously and slowly it pulsated in and out. (It is common experience that in all periodic phenomena, a larger object has a longer period than a smaller one.) Therefore, a more luminous Cepheid has a longer period than a less luminous Cepheid. The new yardstick of the Universe, first brandished by Shapley, turns out to be a logical consequence of differences in stellar properties and not merely a mysteriously lucky break.

was never a star at all. If such a black dwarf happened to form in the neighborhood of a luminous star, it would be a planet. How many black dwarfs there may be in the Universe, which are so far from luminous stars that they cannot be detected, cannot be predicted. Shapley suspects there may be many and one astronomer has even suggested that black dwarfs are the common companions of stars and that ordinary planetary systems like our own are rare.

The exact boundary between a glowing star and a cold black dwarf is not likely to be a sharp one. Some of the larger black dwarfs may work up enough nuclear fusion at the center to be gently warm at the surface.

Even Jupiter, only 1/1000 the mass of the Sun, may not be entirely dead at the center. In 1965, measurements of Jupiter's infrared radiation, made at Catalina Station near Tucson, showed it to be radiating perhaps 2.5 times as much heat as it receives from the Sun. Its surface, according to calculations at Catalina, should be — 170° C. if the Sun were Jupiter's only source of heat. Instead, the surface temperature is — 145° C., twenty-five degrees higher. The extra heat may be derived from the heat of compression at Jupiter's center, and Jupiter may thus, after a fashion, be considered a particularly tiny and particularly cold star.

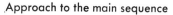

Approach to the main sequence

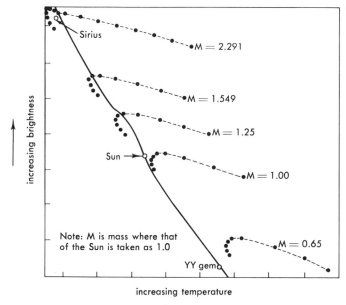

Let us, however, return now to ordinary stars that are hot enough to glow brightly.

The initial stage of stellar evolution, in which a loose aggregation of dust and gas compresses itself and travels toward some part of the main sequence, produces energy chiefly from the gravitational field, the process Helmholtz had pictured to be taking place throughout a star's lifetime. The gravitational source is not, however, a very large one in individual stars, and this stage passes comparatively quickly, in no more than half a million years perhaps. In no time at all (cosmically speaking), the star has reached the main sequence, where the central temperature reaches the point of igniting hydrogen fusion, which then serves as the main source of energy.

This supply of energy is a vast and steady one. In most stars, hydrogen fusion suffices to supply energy at a virtually constant rate over long periods of time. While this is so, the star does not move on the main sequence to any large extent. Any considerable movement, up or down, would represent an upset of the delicate equilibrium between gravitation and temperature.

If our Sun, for instance, were to heat up, for some reason, to the point where its surface temperature was 30,000° C., so that it had suddenly moved to the upper left end of the main sequence, the outward pressure would become so enormously much higher than the inward gravitational compression that it would explode. Only a star of relatively enormous massiveness could possess the gravitational field required to hold its structure together under the explosive force of the outward pressure produced by such high temperature. The Sun is not massive enough for this purpose and, in all likelihood, it never was and never will be. The thought therefore that it was once a Class O star and slid down the main sequence through B, A, and F to its present position as a class G star is untenable. Its present position on the main sequence is the only position, thanks to its mass, that the Sun can comfortably occupy.

Stars remain on the main sequence (more or less motionlessly, in whatever position their mass dictates) for so large a percentage of their total lifetime that more than 99 percent of the stars we can see are on it. In other words, there is less than one chance in a hundred that, in observing a particular star, we will catch it during the relatively short interval before it has reached the main sequence or after it has left it.

Still, some stars remain on the main sequence longer than do others. A large star has a greater fuel supply than a small star, but it must maintain itself at a higher temperature and therefore must consume its fuel at a more rapid rate.

From the mass-luminosity relationship, it can be shown that the rate of fuel consumption increases much more rapidly with mass than the fuel supply does. The larger and hotter a star, therefore, the shorter a time its fuel will last, and the shorter a time it will remain on the main sequence.

Our Sun, for instance, of spectral class G, will stay on the main sequence for a total of some 13 eons. Of these, 5 eons have passed and about 8 eons remain, showing that the Sun is still rather less than middle-aged. A class F star, a little hotter and larger than the Sun, may have more hydrogen to begin with, but consumes it at a sufficiently faster rate to allow it to be on main sequence only 4 to 10 eons. In general, the larger and hotter a star, the briefer its stay on the main sequence. If the Sun were a class A star and had been on the main sequence as long as it has, it would be getting ready to move off it now.

The hottest stars of all burn their relatively tremendous supply of fuel at so rapid a rate that their stay on the main sequence is measured not in eons but in mere tens and hundreds of millions of years. The most luminous star, S Doradus (see page 80), can stay on the main sequence only two or three million years.

If we consider the period before arrival at the main sequence as being short enough to ignore, we can summarize Eddington's view by making the general rule that the brighter a star the shorter-lived it must be. (This is just the reverse of the conclusion one might arrive at from the slide-theory.) The red dwarfs are not near extinction but may continue shining as they do, stingily doling out their small supply of hydrogen, for many eons after the Sun has reached its end. On the other hand, the huge bright stars are not at the beginning of their cycles at all but are expending their fuel so prodigally that they will reach their ends while our Sun is still plodding along exactly as it is doing now.

Eddington's mass-luminosity relation produced a rather surprising by-product that also had its effect on views of stellar evolution. He had worked out the relationship on the assumption that stars possessed the properties of gases throughout their structure (see page 121). At first, he accepted the general view of the early 1920's that only the red giants were gaseous throughout and that the stars of the main sequence, and particularly the dwarf stars, had nongaseous cores. He expected, therefore, that his conclusions would not hold for stars of the main sequence.

To his surprise, though, whenever his conclusions could be tested by observation, they were found to hold for all stars, dwarfs as well as giants. He had to conclude that all stars, the dwarfs included, were gaseous throughout — a conclusion still firmly accepted today.

This result further damaged the original slide-theory of evolution, for it made it difficult to explain why a star should slide down the main sequence, contracting and yet cooling down. The gas laws made it necessary to suppose that contraction would be accompanied by heating, not cooling.

By the mid-1920's, then, the slide-theory of stellar evolution was dead, and what we might call the modern theory was established.

Interstellar Gas

So short are the lifetimes of the brightest stars that they could not have existed in their present form when the dinosaurs ranged the Earth. On a cosmic time-scale they are ephemeral. It is no surprise, then, that so few exist at any one time and that so few can be seen in the heavens today. It is estimated that only one star in 20 million is as luminous as Rigel. Even so, there must be 6000 of them in the Galaxy altogether, and, considering their short life span, all these must have been formed within the last few million years. It may be that every 500 years, on the average, such a star is formed.

In that case, stars must surely have been coming into existence in historic times and even, perhaps, right now. Is it not possible that we may actually see stars in the process of formation if we look carefully?

It is hard to tell whether this is so because the process is so slow in terms of human lifetimes (even if rapid in terms of stellar lifetimes) that over the short period of detailed observations, results are not clear-cut. Furthermore, stars in the formation-stage are not easily visible. There are objects in certain nebulae that might be taken to represent stars in the process of formation. In the Rosette Nebula there are many dark globules that might conceivably be matter condensing on the way toward the main sequence. Other suspected sites of present-day star formation include the Orion Nebula and the Nebula NGC 6611 in the constellation Serpens.

There is even a whole class of objects which may be stars that have only recently (in astronomic terms) begun to shine. These are called "T Tauri variables" because the first one to be detected was T Tauri, by the American astronomer, Alfred Harrison Joy in the 1940's. It is even possible that at least one case is known where a mass of gas condensed, reached ignition point, and began to shine as a star under the very eyes of astronomers. In 1936, the star FU Orionis appeared where no star had been before and has been shining steadily ever since. (If it had been a nova it would have faded away long ago.)

But out of what can the new stars be formed?

Astronomers generally agree that stars are, in the beginning, vast clouds of gas and dust; eons ago when the Galaxy was being formed, this raw material of the stars must have been plentiful. The Galaxy itself must have been nothing more than a huge mass of swirling matter out of which eddies separated and condensed to form stars. But now, with a hundred billion or more stars already condensed out of the primordial swirl, how much raw material can be left?

I have already mentioned the existence of interstellar dust, collecting in places in sufficient quantity to block off starlight as dark nebulae, or reflecting starlight to become luminous nebulae. There is also interstellar dust more generally spread through interstellar space and scattering and dimming starlight generally (see page 75). The effect is important, but it does not take much dust to turn the trick, and that dust alone is not present in sufficient quantities to serve as a generous store for star-formation.

More important is the existence of interstellar gas. Individual atoms or molecules of gases do not absorb or scatter light efficiently and therefore do not make themselves evident as clearly as dust would, even though the gas might be present in far greater quantity.

Individual gas atoms will, however, absorb specific wavelengths of light, just as the atoms in the Solar atmosphere will. The concentration of the gas in interstellar space must be so thin that light absorption over ordinary distances can only be vanishingly small and quite immeasurable. Over hundreds and thousands of light-years, however, cumulative absorption would pile up to measurable levels. It might be possible, therefore, that some lines in stellar spectra could not result from the gases immediately surrounding the stars, but from the very thin gas spread out all the way between the stars and ourselves.

The first indication of this came about through spectroscopic studies of the binary stars. Some binaries revolve about their center of gravity in a plane that is placed edgewise or nearly edgewise to us. When both stars are luminous, the eclipse of one by another is not very effective in changing the quantity of light reaching us, and it is difficult to detect such binaries if the separate bodies are too close to be seen individually by telescope.

However, as the bodies revolve in a plane edgewise to us, one of the pair will be receding from us, while the other is advancing toward us. After a while, one will move behind the other and both will be moving transverse to the line of sight, one to the right and one to the left. Later still, the one that had earlier been receding will now be advancing toward us, while the other, which had been advancing is now receding. They cross transversely again and then the process begins again.

When the components of the binary are moving in such a way that one is approaching and the other receding, the spectral lines belonging to the first will shift toward the violet and those belonging to the other will shift toward the red. When the components are both traveling transversely, there will be no shift in either direction for either star. If the spectra of the two stars are of the same class then, during the transverse-motion stage, the spectral lines of the two will coincide. During the approach-and-recede stage, however, the lines will become double as one set shifts in one direction and the second in the other. In the

Spectroscopic binaries

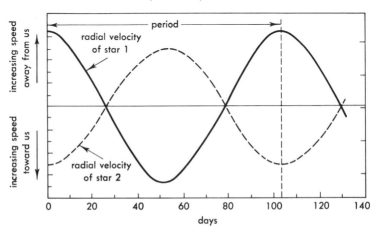

course of the rotation, the spectral lines will double twice each period.

Indeed, it is possible through the behavior of the spectral lines to recognize a star as a binary even when all optical evidence of the fact is lacking. In 1889, the American astronomer Antonia C. Maury (1866–1952) noted such a periodic doubling of lines in the case of Mizar, one of the stars in the handle of the Big Dipper. This was the first star to be recognized as a "spectroscopic binary." Several thousand cases have been discovered since and it may be that one star in a thousand is such a close-whirling pair—so close their atmospheres overlap.

In 1904, the German astronomer Johannes Franz Hartmann (1865–1936) was studying such a spectroscopic binary, Delta Orionis. He noted that during the periodic doubling of lines, one of the lines did *not* double. Something was absorbing a specific wavelength of light and was not sharing the motion of either component of the double star system. It might have been a very massive third component of the system, a component so massive that the center of gravity of the entire system would be close to its own center and it would scarcely appear to move. Yet if such a massive component were luminous it should be seen; and if it were not, it should indicate its presence by eclipses, as in the case of Algol.

It seemed to Hartmann much more likely that the motionless absorption line was produced by the cumulative effect of excessively thin gas in the space between Delta Orionis and ourselves. Hartmann's conclusion was not immediately accepted, but additional evidence was reported by other astronomers, in particular the Russian-American astronomer Otto Struve (1897–1963) in 1928. Interstellar gas is now accepted as a feature of the Galaxy, and its total mass is perhaps 50 to 100 times that of the dust in the Galaxy.

The motionless spectral line first observed by Hartmann was one that was produced by calcium atoms, so it seemed obvious that the interstellar gas included calcium. Other atoms were also detected, but the composition of the gas could not be determined merely from the spectral lines. A gas that happened to absorb certain wavelengths in the visible light region very strongly — as calcium does — might impress its mark on the spectrum even though it was present in only minor quantities. By the 1950's it became quite plain that the preponderant component of the interstellar gas was the much less obtrusive (spectroscopically speaking) hydrogen.

It is estimated today that 90 percent of all the atoms in the Universe are hydrogen, the simplest of all the atoms, and that 9 percent are helium, the next simplest. All other types of atoms make up the remaining 1 percent. In short, the elemental makeup of the Sun is thought to be rather typical of the elemental makeup of the Universe generally.

If the interstellar gas is mainly hydrogen and helium, of what does the dust consist? Helium has virtually no tendency to clump together, and hydrogen forms two-atom molecules which also have little tendency to clump together. Dust must form, therefore, with the aid of some of the minor constituents of interstellar matter, but not a constituent that is *too* minor, for the dust content of the Galaxy is substantial.

One suggestion focusses on oxygen, the most common of the minor elements. An oxygen atom can easily combine with a hydrogen atom to form a "hydroxyl group," and in 1963 these were actually detected in the interstellar matter. An oxygen atom can also combine with two hydrogen atoms to form a water molecule, and water molecules have a srong tendency to clump together. The interstellar dust might, therefore, be made up of ice crystals in large part. It seems likely there may also be granules of rocky materials called "silicates," and even tiny metallic particles, though these have not yet been detected.

What has been detected, rather to the surprise of astronomers, are molecules made up of three or more atoms, other than water itself. These were detected by the fact that they radiated specific wavelengths of light-like radiations called "radio waves," which will be discussed later in the book.

In 1969, a four-atom combination was discovered. It was the formaldehyde molecule made up of a carbon atom, an oxygen atom, and two hydrogen atoms. In 1970, the five-atom combination of cyanoacetylene (three carbon atoms, a hydrogen atom, and a nitrogen atom) was detected. In 1971, a seven-atom combination, methylacetylene (three carbon atoms and four hydrogen atoms) was detected. By the end of the 1970's, some thirty different molecules, almost all of them involving the carbon atom, have been detected.

It is clear now that the interstellar dust clouds have a complicated chemistry. Astronomers are not yet clear just how such molecules may have formed in the vast emptiness of space, but they are there and it is interesting that the carbon atom is so involved. It is the carbon atom, after all, that is central to the kind of molecules that make up living tissue.

Nevertheless, all these molecules make up a small proportion of the interstellar gas and dust. These remain mostly hydrogen and helium and, though thinly spread out, occupy an immense volume, and amount, in total, to a large mass. Some estimates have made the interstellar matter of the Galaxy equal in mass to all the stars, but this is almost certainly an overestimate. The most recent determinations would make the mass of interstellar gas only 2 percent that of the stars, although the spiral arms would be considerably more gassy than the Galactic nucleus. In the spiral arms the mass of interstellar material might be as high as 10 to 15 percent that of the stars.

Even at the lowest estimate, there would be enough interstellar matter in a galaxy like ours to make up two or three billion stars if it were all collected, so it is no surprise that some stars may be forming even today out of this largely hydrogenous interstellar matter, or that some stars formed one to ten million years ago and shine today with supernal brightness.

Some other galaxies may serve as even richer sources of new star material than ours does. The Large Magellanic Cloud, for instance, has perhaps three times the concentration of interstellar gas that the Galaxy does.

An explanation can now be offered for the unexpected hydrogen deficit in the Sun and for its apparently too-large supply of helium, as well as for the fact that a planet such as the Earth is made up almost entirely of elements more complicated than helium. Apparently the gaseous material out of which the Solar system formed already contained a considerable supply of helium and a small quantity of more complex atoms as well.

The question then is: Where did the helium and the more complex atoms in the interstellar gas come from?

We might suppose that the gas out of which the Galaxy was formed simply contained a quantity of helium and more complicated atoms to begin with. It is much more tempting, however, to suppose that only hydrogen, the simplest of all atoms, was present to begin with and that all other atoms formed from hydrogen. Yet the only processes by which hydrogen will fuse to form other atoms, as far as we know, require the conditions present in the cores of stars. In that case, how did the helium and other atoms get back into the gas?

Let us keep that question in mind as we continue to consider the possible course of stellar evolution.

Beyond the Main Sequence

What happens to a star after it consumes so much fuel that it can no longer maintain the balance between gravitation and temperature and can, therefore, no longer remain on the main sequence?

The answer to this arises partly out of theory and partly out of observation. Nuclear physicists can work out elaborate theories of what might happen in the interior of stars under certain conditions of temperature, pressure, and chemical composition, and this can be checked against what we observe in the heavens.

Astronomers can, for instance, observe those star-clusters, such as the Pleiades, which are close enough to us so that the individual members can be studied spectroscopically. The stars in such a cluster are all at about the same distance from us, so that their order of apparent brightness is identical with their order of luminosity. That gives us the vertical axis of the H-R diagram. If the individual spectra are studied, we have the horizontal axis also. We can, in short, make a small and special H-R diagram of the cluster stars only.

Furthermore, we can assume that all the stars in such a cluster are the same chronological age. It seems reasonable to suppose that a large volume of gas spreading out over the entire volume of the cluster at one time in the distant past condensed into the various stars making up the cluster over so short a time period, cosmically speaking, that all the stars of the cluster can be considered as having been born at once. If this is not so and if the stars of the cluster came into being independently, then it is hard to explain their close association today.

If chronological age were the only criterion of stellar evolution, then all the stars in a cluster, being of the same age, ought to be at the same stage of evolution. All would be at some one point on the main sequence, or at some one point before reaching it, or at some one point after leaving it.

Chronological age is not, however, the only criterion. Mass is another. The stars in a cluster vary over a range of mass, and the more massive a star the more rapidly it develops. A massive star born at a particular time would have reached a later stage in evolution than a less massive star born at the same period. In a cluster of stars, then, the least massive stars would be at the earliest stage of the evolutionary cycle, and successively more massive stars would be at later and later stages. If each star is placed at its approximate point in the H-R diagram, the entire course of evolution is sketched out, up to the point of farthest evolution in the cluster.

Trumpler (the man who demonstrated the existence of interstellar dust) carried on such systematic studies of clusters in 1925, and similar observations in great detail have been made ever since. Combining these observations with theory, astronomers now feel confident they can describe the evolutionary adventures of an individual star.

1. A mass of gas and dust contracts and heats until it reaches the main sequence. This will take perhaps 100,000 years for a large mass of gas which will end at the hot upper end of the main sequence because the relatively large gravitational field of a large mass will cause it to contract relatively quickly. A star the mass of our Sun will contract to main sequence in perhaps 2 million years.

2. Stars will then stay on the main sequence anywhere from millions of years to tens of eons, depending on their mass. While a star is on the main sequence, nuclear reactions are proceeding in the high temperature core, which gradually loses hydrogen and accumulates helium. Thus, calculations reported in 1966 indicate that the Sun's core contains six helium atoms for every hydrogen atom, while its outskirts see the ratio just about reversed, seven hydrogen atoms for every helium atom.

3. When the core reaches some critical content of little hydrogen and much helium, the star begins to expand and, consequently, cool. It then leaves the main sequence and, on the H-R diagram, can be pictured as moving upward and to the right. In some cases, it is thought, the star reaches a Cepheid stage, pulsating regularly for some millions of years. In other cases, the expansion continues more or less smoothly until the star is enormous and its matter, in the outer layers at least, very rarefied. It has become a red giant which thus represents a late stage in stellar evolution (rather than an early one as was suggested in the slide-theory of stellar evolution).

Our Sun, after about 8 eons, will begin to expand its way to the red giant stage, and if mankind has not left the Earth by then (or put an end to itself long before) that will be our finish.

Naturally, the larger and more massive a star, the more tremendous a red giant it will balloon into. The red giant into which our Sun will someday bloat will not be a particularly impressive specimen of the class. Red giants such as Betelgeuse and Antares developed out of main sequence stars considerably more massive than the Sun. The most massive stars may have grown still larger and cooler to form the infrared giants.

4. By the time the red giant stage has been fully reached, the hydrogen in the core is entirely consumed. The core has been constantly increasing in mass as more and more helium collects in it. While the core itself is contracting and growing hotter, the outer layers are expanding because of increasing temperature. Eventually, a core temperature of about 140,000,000° C. (nearly ten

Stellar evolution

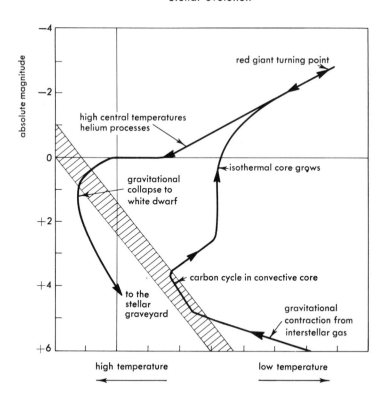

times the temperature of the Sun's interior at present) is reached, and that is hot enough to ignite a nuclear reaction in which three helium atomic nuclei combine to form the atomic nucleus of a carbon atom. (This is sometimes called the "Salpeter process" after the Austro-American physicist, Edwin Ernest Salpeter (1924–), who first worked out the process.) Such "helium-burning stars" begin to contract and heat up again.

The development of helium fusion does not mean that the star has gained a new lease on life comparable to the early stage of hydrogen fusion. Helium fusion does not supply as much energy as hydrogen fusion does.

Suppose we consider twelve hydrogen-1 nuclei fusing to three helium-4 nuclei, and then the three helium-4 nuclei fusing into a single carbon-12 nucleus. The twelve hydrogen-1 nuclei have a total mass of 12.0956; the three helium-4 nuclei a total mass of 12.0078; and the carbon-12 nucleus a mass of 12,000. The mass-loss (and, therefore, energy release) in the hydrogen-fusion stage is 0.0878, while that in the helium-fusion stage is only 0.0078. Thus, mass for mass, helium fusion will yield only 9 percent as much energy as hydrogen fusion.

The star can make do, however, by fusing carbon atoms still further to form more complicated nuclei, but this cannot continue forever. A dead end is reached with iron. The nuclei of the iron atoms are the most stable of all. Once the iron nucleus is reached, no more energy can be obtained out of nuclear reactions involving it. Whether the iron nucleus is built up to more complicated ones or broken down to less complicated ones, energy is *not* released; energy must be supplied.

And, as it happens, little energy can be obtained even if nuclei are followed all the way up to the dead end, once hydrogen fusion is over and done with. Imagine 56 hydrogen-1 nuclei converted to 14 helium-4 nuclei, and that these 14 helium-4 nuclei are then converted into a single iron-56 nucleus. The respective masses are 56.4463, 56.0264, and 55.9349. The mass-loss in passing from hydrogen to helium is 0.4199, and from helium to iron is 0.0915. The energy gained from the conversion of helium all the way to iron is only 22 percent that of the conversion of hydrogen to helium, weight for weight. We might say, then, that when a star has consumed its hydrogen, its life as a nuclear reactor, is four-fifths over.

After the red giant stage, as a star shrinks and heats up steadily, a core forms within the core and another core within it, and so on. Each core as it forms contains more complicated atoms than the one before, until finally an end comes with an innermost core rich in iron.

This process of shrinking and heating may be pictured on the H-R diagram as a rapid passing leftward and downward. The star reaches and crosses the main sequence heading for the lower left region of the diagram, a region that would contain hot stars of low luminosity.

The lower left region is, naturally, just the opposite of the upper right region where cool stars of great luminosity (the red giants) are to be found. Just as cool stars can be highly luminous if they are very large, so hot stars can be unusually dim if they are very small.

In order to understand what happens to stars when their nuclear reactions approach an end, we must consider the nature of very hot, very small stars—a group I have not yet discussed.

11

Stellar Explosions

White Dwarfs

Knowledge concerning the hot but dim stars began with Bessel's discovery of the companion of Sirius in 1844 (see page 39). Sirius and its companion moved in elliptical orbits about their mutual center of gravity, with a period of revolution of fifty years. Sirius is sometimes referred to as "Sirius A," therefore, in deference to its existence as part of a binary system, while its smaller companion is "Sirius B."

Any object capable of swinging a star like Sirius in an orbit large enough to be visible from Earth must itself produce a star-sized gravitational field. In fact, it could be calculated that Sirius' companion had to have a mass fully half that of Sirius A and must therefore be about as massive as our Sun. Yet Sirius B could not be seen, so that Bessel assumed it to be merely the burnt-out cinder of a star.

That same year, Bessel reported a similar dark companion for Procyon so that one could speak of "Procyon A" and "Procyon B." It seemed very likely that such dim stars might be quite common. It was thought that only their lack of light prevented them from being seen; and that they could be detected only when they were part of a multiple star system and their gravitational fields could alter the motions of a visible star to such an extent that the results could be made out from Earth.

There was sufficient faith in this notion so that in 1851, the German astronomer, Christian August Friedrich Peters (1806–1880), calculated the orbit the invisible companion of Sirius would have from the gathered data on the waves that Sirius A was displaying in its motion.

To be sure, neither companion turned out, in the end, to be completely dark. In 1862, Clark detected Sirius B as a star of magnitude 7.1. This is not terribly dim, of course, but the Sirius star system is only 8.8 light-years away. To be at that distance and still appear so dim, Sirius B had to have a luminosity only a hundredth that of our Sun, or less. In 1895, the German-American astronomer John Martin Schaeberle (1853–1924) detected Procyon B and that turned out to be only of the eleventh magnitude. Even allowing for the fact that Procyon B was a bit farther from us than Sirius B, it could quickly be shown that Procyon B was even dimmer than Sirius B. Such stars might not be completely dark, but they were certainly dwarf stars.

It was taken rather for granted at the turn of the century that stars like Sirius B and Procyon B were dying stars that were dim primarily because the stellar fires were flickering out. One might suppose that such stars might fit neatly among the red dwarfs at the tail of the main sequence.

However, even as the H-R diagram was being worked out, it became quite apparent that Sirius B, for instance, would not fit in just that place. To be at the tail end of the main sequence, a star would have to be very cool, and, therefore, a deep red in color. Sirius B, however, was not red. It shone with a clear white light. If it was a dwarf, it was a "white dwarf."

In 1914, the American astronomer Walter Sydney Adams (1876–1956) succeeded in taking the spectrum of Sirius B and found it to be a spectral class A star, exactly as Sirius A itself was. This meant that Sirius B had to have a surface temperature as high as that of Sirius A (10,000°C.) and higher than the mere 6000° C. surface temperature of the Sun.

But if Sirius B was hotter than the Sun, it should have a surface brighter than the Sun, square mile for square mile. The fact that Sirius B was so much less luminous than the Sun could only mean that Sirius B possessed very few square miles of surface area. It was a star that was white-hot but very small, just the kind of star you would expect to find in the lower left region of the H-R diagram, the region mentioned at the end of the previous chapter.

In fact, it would have to be *quite* small. To account for its dimness, Sirius B would have to have a diameter of not more than 7,000 miles, according to the latest figures, and be rather smaller in size than the planet Earth. It would be a white dwarf indeed.

And yet Sirius B would still have the mass of the Sun. That was determined from the gravitational effect of Sirius B on Sirius A, and it could not be argued away. For a star, then, to be as small as Uranus and as massive as the Sun brought up serious questions of density—questions that would have raised insuperable difficulties in the nineteenth century but, as it turned out, could be answered in the twentieth.

Indeed, the problem of stellar densities arose not only about unusual stars such as Sirius B, but about our own Sun.

Once the distance of the Sun was known, it was easy to calculate its real diameter, and therefore its volume, from its apparent diameter. Its volume turned out to be 1,300,000 times that of the Earth. Its mass, calculated from its gravitational effect on the Earth, turned out to be 333,500 times the mass of the Earth.

Density is determined by dividing the mass of an object by its volume. The Sun's density should therefore be $333,500 \div 1,300,000$ or just a little over one-fourth that of the Earth. The density of the Earth is 5.5 grams per cubic centimeter and that of the Sun is therefore 1.41 grams per cubic centimeter.

The density of water is 1.00 grams per cubic centimeter, so the density of the Sun is 1.41 times that of water. Moreover, this figure represents an overall average. Certainly the density of the Sun's outermost layers must be considerably less than that, and the density of its interior core (under the vast pressure of the weight of those outermost layers) must be considerably more than 1.41 to compensate.

There is an analogy here in the case of the Earth. The overall density of the Earth, as I have said, is 5.5 grams per cubic centimeter, but the outermost rocky crust has a density of only 2 6 grams per cubic centimeter, while the density rises at the Earth's center to 11.5 grams per cubic centimeter.

The change in density is much greater for the Sun than for the Earth, since pressures are much higher at the center of the huge Sun than at the center of the relatively small Earth.

Once Eddington began to explore the internal structure of the Sun and other stars, the figures for central densities were found to be unbelievably high. Calculations show that in order for the Sun to remain in gravity-temperature balance, the density at its center may rise as high as 100 grams per cubic centimeter. This is five times as dense as platinum (or its sister elements, iridium and osmium), and these are the densest substances known on Earth.

Furthermore, while stars that are larger and hotter than the Sun are less dense, and giants like Epsilon Aurigae are very rarefied indeed, stars that are smaller and dimmer than the Sun are more dense. A red dwarf, like the star known as Kruger 60B, has a mass that is one-fifth that of the Sun. Its volume, however, is only $1/125$ of the Sun and its density is $1/5 \div 1/125$ or twenty-five times that of the Sun. Its average density must therefore be 35 grams per cubic centimeter, which is more than half again as dense as platinum, and its density at the center must be hundreds of times that of platinum.

In the first decades of the twentieth century, when these figures were not known exactly but when they were expected to be quite high, it is not surprising that it was assumed that the gas in stellar interiors was compressed to the point where it no longer acted as a true gas. Theories concerning stellar energies and stellar evolutions were based on the assumption of nongaseous cores— and proved to be all wrong.

Eddington's work in the 1920's showed that all stars, down to the red dwarfs, behaved as though they were perfectly gaseous throughout, despite their densities, since they obeyed the mass-luminosity relation which was postulated on the assumption of all-gas stars.

But how can substances of such high density behave as though they were thin gases? Indeed, how can substances of such high density exist at all? If atoms were, as nineteenth-century chemists believed, hard little billiard balls, unbreakable and incompressible, such high densities would not be possible. In ordinary solids here on Earth, such atoms are already in contact, and the density of the platinum metals would represent very nearly the maximum possible.

From the 1890's on, however, all notions about atomic structure were revolutionized. It became clearer and clearer that the atoms were not featureless little billiard balls, but were complex structures made up of "subatomic particles" that were individually far tinier than the intact atom. Whereas the atom as a whole had a diameter of the order of a hundred-millionth of a centimeter, the diameter of the subatomic particles was something like a ten-trillionth of a centimeter. To put it in more easily grasped terms, the subatomic particle was only 1/100,000 as wide as the intact atom; it would take 100,000 subatomic particles laid side by side to stretch across the diameter of a single atom.

The volume of an atom was $100,000 \times 100,000 \times 100,000$, or 1,000,000,000,000,000 times that of a single subatomic particle. Since even the most complex atom contains only a little over 300 subatomic particles, it can clearly be seen that the intact atom is largely empty space, held in its wide-open structure by the electromagnetic forces that kept a few electrons moving through wide spaces about the tiny atomic nucleus at the center of the atom. If an atom could be broken up into its individual subatomic particles and compressed, the whole system could be made to shrink to a tiny fraction of its former self. (As an analogy, think first of the amount of space it would take to store a dozen hatboxes, then imagine the hatboxes torn into small pieces of cardboard and think of how much smaller a space they could be packed into. The case of the atom is far more extreme.)

At high temperatures, the atom is stripped of its outermost particles, the electrons; if the temperature reaches values high enough (and at the center of stars it certainly does), all the electrons are stripped away, leaving naked atomic nuclei. Under the tremendous pressures of a star's interior, the electrons and nuclei can be compressed into a far smaller volume than the original atoms would have taken up. As the volume decreases, the density correspondingly increases to many times that of platinum.

Such crushed-together quantities of subatomic particles are usually referred to as "degenerate matter." There is no question but that there is a core of degenerate matter within the Sun and in fact, within all other stars. This makes sense, too. In intact atoms, the outer electrons completely shield the atomic nuclei and prevent them from colliding directly and combining with each other. It is only when the electrons are stripped away completely that nuclear fusions can take place rapidly enough to support stellar radiation.

Furthermore, although degenerate matter is squeezed into immensely high densities, the individual subatomic particles making it up are so small that the degenerate matter is still very largely empty space. Degenerate matter, composed of separate subatomic particles, is as nearly empty space, in fact, as are the much less dense ordinary gases made up of the much more bulky intact atoms. It is for this reason that degenerate matter may have incredibly high densities and still act like a gas.

Yet the densities of even the red dwarfs are as nothing compared with those of the white dwarfs. Sirius B, with the volume of less than the Earth and the mass of the Sun, must have a density about 530,000 times that of the Earth or 130,000 times that of platinum. A cubic centimeter of the average material of Sirius B would weigh 2,900 kilograms. A cubic inch of it would weigh 52.4 tons. The central regions of Sirius B may be as much as 12 times the average density so that a cubic inch of the core of the star would weigh 630 tons. Yet Sirius B also seems to behave as though it were completely gaseous.

Clearly, Sirius B must be made up almost exclusively of degenerate matter. This was hard to accept even in the 1920's, but corroborative evidence was found. In 1915, Einstein had worked out his general theory of relativity which predicted, among other things, that light would experience a red shift if it traveled outward against a gravitational force. Ordinary gravitational fields, his theory showed, would produce a red shift too small to measure, and at the time he advanced his theory, he did not realize that any sufficiently strong field might exist.

Eddington pointed out that if Sirius B were really as dense as it seemed to be, it ought to have a surface gravity much greater than that of the Sun. The best figures we have today indicate that the surface gravity of Sirius B is, in fact, 16,500 times that of the Sun. Under such conditions, the "Einstein shift" would be measurable. In 1925, W. S. Adams checked the spectrum of Sirius B even more carefully, measured the position of various spectral lines after allowing for the star's radial motion and, sure enough, the Einstein shift was found. This was an important point in favor of the validity of the general theory of relativity. It was also an important point in favor of the super-density of Sirius B.

Sirius B is by no means unique. Other super-dense white dwarfs are known. Procyon B is one of them, of course, with about 0.65 times the mass of the Sun. About 250 such objects have been discovered, many by the Dutch-American astronomer Willem Jacob Luyten (1899–   ). In 1962, he discovered one with a diameter only half that of the Moon, the smallest one yet found. In 1966, the Swiss-American astronomer, Fritz Zwicky (1898–1974), spotted a binary made up of two white dwarfs.

Two hundred fifty white dwarfs does not seem to be a very impressive number against many billions of ordinary stars, but remember their small size and dimness. They can only be seen if they are quite close to us, whereas ordinary stars can often be seen over immense gaps of space. The fact that so many white dwarfs have been found despite the handicap of dimness makes it plain that they must be very common objects indeed. Some have even estimated that they make up 3 percent of all the stars in the Galaxy, which would mean a total of some three billion white dwarfs in the Galaxy.

To return to the matter brought up at the end of the previous chapter, is it toward the white dwarf stage that the stars wih nuclear fuel gone are heading, as they approach the lower left region of the H-R diagram? Apparently, yes—but not always by way of a smooth and peaceful transition.

Supernovae

In the light of the new nuclear view of stars, it would seem tempting to interpret the nova (see page 88) as a kind of Cepheid gone wrong, so to speak. Where the ordinary Cepheid pulsates in a controlled and measured fashion, puffing up and settling back over and over, a nova is a star which, after a long period of quiescence, manages, for some reason, to build up a sudden head of outward pressure and—quite literally—explodes.

The nova's luminosity increases rapidly from 5000 to 100,000 times, as its surface area increases, especially since it is suffering a runaway surge of radiation which keeps it from cooling down as it expands. At its peak, such a nova reaches an absolute magnitude of −8, at which time it is about 200,000 times as bright as the Sun.

This brightness peak only lasts a few days, however. The force of the explosion puffs a portion of the star's matter out into space, and with it goes much of the energy. What is left of the star begins to collapse again, like a punctured balloon, and the star dims. It takes several months, perhaps, to get back to its pre-nova brightness and after that it continues much as it did before.

Such an explosion is catastrophic on an earthly scale. If it happened to the Sun, Earth's oceans would boil and its life would probably be wiped out. On a stellar scale, the nova explosion is not too bad. The peak brightness is high but, even at its height, a nova is not as bright as S Doradus is all the time. As for the material lost in the course of the nova explosion, this does not amount to more than about 1/100,000 of the mass of the star— scarcely enough to matter.

After a quiescent period, a nova is quite capable of undergoing another explosion with intervals of anywhere from 10 to 100 years. Nova Persei, which exploded in 1901 (see page 90), brightened again in 1966. One star, T Pyxidis, has been observed to have four such novalike peaks of brightness since 1890, and then brightened a fifth time in 1966. Whatever property serving it on its way to the peak and do not catch it—as all too often happens—only after the fact), absorption lines in its spectrum show a strong violet-shift, indicating the star is approaching us. Part of it certainly is, for the exploding outer layer moves rapidly away from the star and the portion between ourselves and the star approaches us at a good, spanking clip.

The explosion, or its aftermath, is even visible optically in some cases. After Nova Aquilae appeared in 1918, E. E. Barnard noticed that it was surrounded by a nebulous sphere that had not been there before. This sphere, presumably composed of exploding gases, continued to move outward at a uniform rate, growing slowly more voluminous and fainter until 1941 when it had become too faint to detect. Other novae exhibited similar phenomena.

The weight of astronomical opinion, however, has now shifted against the view of the nova as simply a more-or-less ordinary star that explodes. A suggestion first proposed in 1955 by Otto Struve has gained favor.

Every nova seems to be a member of a close binary, one of two stars that circle each other at a relatively small distance. The nova itself, when it is quiescent, seems to be a white dwarf; its partner is a large star, some of whose matter can drift away and be caught in the intense gravitational field of the white dwarf. This would happen in an irregular fashion, and every once in a long while some unusual activity in the large star would send a sizable gout of matter toward the white dwarf. This matter would be compressed in the gravitational field of the white dwarf, and this compression would set off a very rapid nuclear reaction releasing enormous energies. This would produce a flash of radiation, and we would become aware of a nova in our sky.

Not all exploding stars are, however, merely novae. This became evident in the mid-1920's when the enormous distance of the Andromeda galaxy first came to be appreciated. If that distance were accepted, what of the nova, S Andromedae, that had appeared in it in 1885 (see page 90)?

When S Andromedae was first detected, it was of the seventh magnitude, but there is a chance that it was not detected until slightly after it had zoomed rapidly upward to its peak, and that it had been visible briefly to the naked eye as a star of slightly better than the sixth magnitude.

This would still make it only a dim star, to be sure, and one that was just barely visible—but to be visible at all to the naked eye at the distance of the Andromeda galaxy bespeaks a brilliance so enormous as to stagger even the hardened astronomer. S Andromedae at its peak outshone all the rest of the Andromeda galaxy. A single star was brighter than the combined brilliance of billions of ordinary stars.

Modern estimates, based on the latest values accepted for the distance of the Andromeda galaxy, make it appear that the absolute magnitude of S Andromedae at its peak was −19. This meant that, for a few days anyway, it was shining with a brilliance equal to a hundred thousand ordinary novae, or nearly ten billion times that of our Sun.

S Andromedae was not merely a nova; it was a "supernova."

Once this was realized, the search was on for other examples of this new and spectacular class of objects. Whereas ordinary novae cannot be seen much beyond the nearest galaxies, the supernova, which is as bright as an entire galaxy, can, of course, be seen as far as galaxies can; that is, as far as our telescopes can reach.

Outstanding in the search for such galaxy-bright supernovae has been Fritz Zwicky who, in the years following 1936, located a number of them in various galaxies. About fifty have been observed all told. Zwicky estimates that whereas ordinary novae appear at the rate of twenty-five per year in any given galaxy, supernovae appear at the rate of only three per thousand years in any given galaxy.

By the end of the 1930's, observations by the German-American astronomers Walter Baade (1893–1960) and Rudolf Leo B. Minkowski (1895–1976) had shown that supernovae could be divided into two varieties, Types I and II, with, possibly, a Type III as well. The Type II supernovae are the less luminous, only a couple of hundred times brighter than an ordinary nova and probably the more numerous, although fewer are seen simply because of their lower luminosity. Type III supernovae are like the Type II, but with a light-curve that fades off more gradually. Type I supernovae are the true giants of the class, and S Andromedae was a Type I supernova as was the supernova of 1054 that gave rise to the Crab Nebula.

Whereas ordinary novae owe their increase in brightness, in

all probability, to flashes of fusion in matter gained from a nearby star, supernovae are indeed explosions and extraordinarily cata-strophic ones. A Type II supernova loses from $1/100$ to $1/10$ of its mass in the course of this explosion, and a giant Type I super-nova loses from $1/10$ to $9/10$ of its mass.

Although a number of novae have been studied within our Galaxy since the invention of the telescope, not one object that can clearly be labeled a supernova has been detected in the three and a half centuries since Galileo first pointed his magnifying tube toward the heavens.

However, looking back into history, there seem to have been three supernovae in our Galaxy in the course of the last thousand years (right up to par, according to Zwicky's estimate). They were the "new stars" of 1054, 1572, and 1604 (see page 170). There are also signs (according to painstaking searches through Oriental records, reported in 1966) that a fourth supernova may have blazed out in 1006.

Of these, the first was the most poorly observed but has turned out to be the most interesting by far. It may have been one of the brightest supernovae that ever formed and possibly the closest in historic times. Moreover, it is unique among supernovae in that it has left behind a remarkable remnant of itself, in the form of what seems, in a small telescope, to be a cloudy patch.

In 1731, the English astronomer John Bevis (1693–1771), reported its existence and in 1758, Messier observed this patch in the constellation Taurus, and entered it in his list of nebulosities. Indeed, it was the first object in his list and is sometimes called M1.

Eighty years later, in 1844, Lord Rosse observed it more closely with his large telescope and was able to make out its struc-ture. It was like nothing else in the heavens, a mass of clearly turbulent gas with numerous filaments of light within it. Because the numerous ragged filaments reminded Rosse of the legs of a crab, he named the object the "Crab Nebula," and that remains its name to this day.

The Crab Nebula looks for all the world like a vast explosion caught in mid-expansion. It takes no imagination at all to see that; the conclusion forces itself on any observer. The fact that it is lo-cated in just about the spot where the Chinese astronomers of 1054 had placed their supernova made it very tempting to consider the Crab Nebula the remnant of that explosion, and by the mid-1920's, when astronomers came to understand the fact that such supernovae existed, this view was generally accepted.

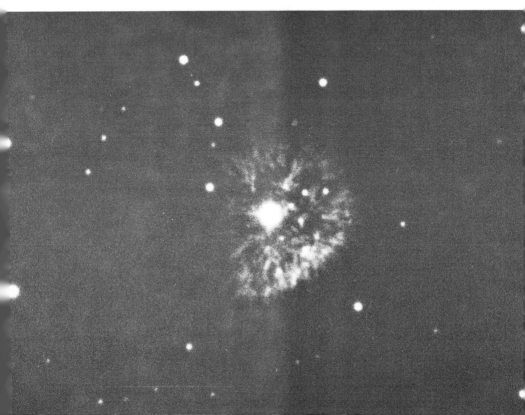

Spheroidal Galaxy. *(Photograph from the Mount Wilson and Palomar Observatories.)*

Nebula About Nova Persei. *(Photograph from the Mount Wilson and Palomar Observatories.)*

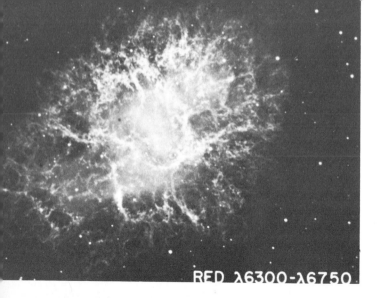

RED λ6300-λ6750

INFRARED λ7200-λ8400

The Crab Nebula—photographed at various wavelengths. *(Photograph from the Mount Wilson and Palomar Observatories.)*

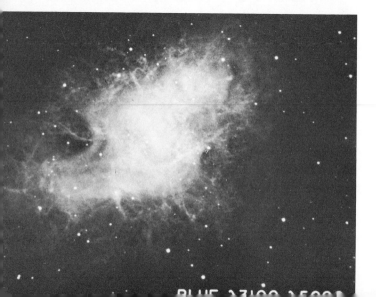

BLUE λ3100-λ5800

Planetary Nebulae. *(Photograph from the Mount Wilson and Palomar Observatories.)*

Cluster of Galaxies. *(Photograph from the Mount Wilson and Palomar Observatories.)*

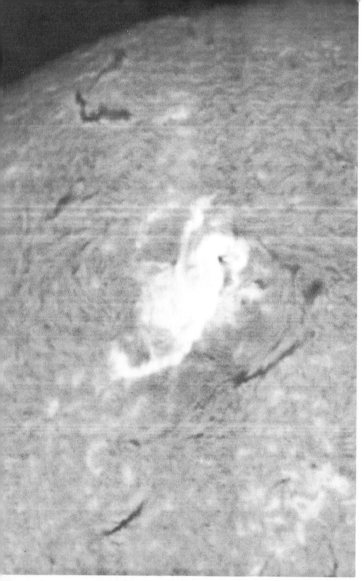

Solar Flare. *(Photograph from the Mount Wilson and Palomar Observatories.)*

Sun's Corona. *(Lick Observatory Photograph.)*

In 1921, J. C. Duncan pointed out that photographs of the Crab Nebula taken over the last generation or so have shown the turbulent gases to be moving outward by an amount that is tiny in terms of seconds of arc per year, but measurable. Spectroscopic evidence also shows that the portions of gas on the side toward us are approaching us at a rate of about 1300 kilometers per second. If the Crab Nebula is really an exploding volume of gas, that gas ought to be moving outward in all directions at roughly equal velocities. The radial velocity in kilometers per second can therefore be set equal to the transverse velocity in seconds of arc per year. This distance at which such a velocity will make itself evident as so many seconds of arc per year can be calculated, and it turns out that the Crab Nebula is about 4500 light-years from us and that the exploding shell of gases now has an extreme diameter of about 6 light-years.

If we take the rate at which the gas is expanding and calculate that backward, we can show that the gas was at its central starting point some 900 years ago, which is exactly what would be expected if the Crab Nebula were indeed the exploding supernova of 1054. By 1942, there seemed no doubt about the matter, thanks chiefly to the careful work of Oort.

No such obvious markers are to be found of Tycho's supernova of 1572 or of Kepler's supernova of 1604 (the latter perhaps a Type II supernova only), although certain faint wisps of gas about 11,400 light-years from us (three times the distance of the Crab Nebula) were identified in 1966 as possible remnants of Tycho's supernova. Again, the Veil Nebula in Cygnus seems to be the remains of a supernova explosion that took place 100,000 or more years ago.

Dying Stars

The two types of objects discussed so far in this chapter—small, very dense stars and exploding stars—may be closely related.

To see how, let us return to stars at the final stage of nuclear fuel consumption—stars that have been accumulating iron in their innermost core and have nowhere else to go in the realm of nuclear reactions.

For such a star to continue to radiate, the one remaining energy source, the gravitational field, must be tapped. Once again the star must contract, as it had done once in the days long past, before nuclear reactions had been ignited in its core. Only now, when the star is radiating energy at a tremendous rate, the contraction must be rapid indeed if enough energy is to be supplied.

Once the nature of degenerate matter was understood, astronomers, could see that this contraction could be exceedingly rapid, and that what had previously been an ordinary star could be converted in almost no time into a tiny white dwarf. The heat of compression would make it white hot, but because of its small surface area, it would be radiating far less energy, altogether, after compression than before. The amount of energy it would be radiating as a white dwarf was small enough to be fed over long eons by an exceeding slow rate of further compression. The process which Helmholtz had once thought applied to all stars generally, turned out to be applicable, indeed, to white dwarfs.

Nor is a white dwarf deprived, by its enormous density, of the possibility of further compression. Much capacity in that direction remains. Sirius B might have a density 125,000 times that of the Sun, but the subatomic particles that swarm about in its largely degenerate structure are by no means in contact. Sirius B would have to shrink to a diameter of a mere eight miles or so (!) before that would be true.

As a white dwarf shrinks, it cools. Its surface temperature may be as high as 50,000° C. at the moment of formation, but Sirius B with its 10,000° C. surface temperature is still a fairly young white dwarf, close to the possibly maximum luminosity of 1/100 that of the Sun. Older white dwarfs are correspondingly cooler, and one, called "van Maanen 2," which has been in the white dwarf state for at least 4 eons, has a surface temperature of only 4000° C. It is distinctly reddish in color so that it is that apparent contradiction in terms, a "red white-dwarf." Still even van Maanen 2 can continue to exist for many, many eons, doling out its gravitational supply of energy, before it flickers out. Such is the extended lifetime of a white dwarf that it may well be that the Galaxy itself is not old enough, yet, to have witnessed the final flicker of a single such star.

But if white dwarfs are dying stars, an interesting problem is raised by the Sirius system. Sirius A and Sirius B must have been formed simultaneously, and yet Sirius A is in the prime of its life while Sirius B is in extreme old age. Why should that be?

If both are of equal chronological age, it can only be that Sirius B lived faster than Sirius A, and this would imply by the mass-luminosity relation that Sirius B is more massive than Sirius A, perhaps far more massive. Yet, Sirius B, in its white dwarf state at least, has only half the mass of Sirius A. What happened to the rest of the mass that must once have belonged to it?

The likeliest possibility is that Sirius B became a red giant, and leaked quantities of matter outward that were captured by the then-smaller Sirius A. By the time Sirius B completed its stay in the red-giant portion of its evolution, and shrank down to a white dwarf, it was considerably less massive than it had been. Sirius A had become considerably more massive, brightened, heated up, and became blue-white in color. It became as we see it now, the more massive of the pair, and its lifetime has been enormously shortened.

Once a star of average size, like our Sun (or as Sirius B became after losing much of its mass to Sirius A), has exhausted the nuclear fuel of its core, it would contract more or less steadily, leaving behind, perhaps, the thin outermost portions of its swollen red-giant body.

If that were the case, we might expect to see freshly made white dwarfs surrounded by a distant halo of thin matter, and we do. There are over a thousand stars we can see that are surrounded by shells of thin gas. We see through the greatest thickness of this gaseous halo at the edges, so that it appears to be a "smoke ring" about the central star. The best known example is the Ring Nebula in the constellation Lyra. Such a ring of gas is reminiscent of Laplace's picture of shells of gas being given off by a rotating nebula that, eventually, was supposed to condense into planets. For that reason, objects of this sort are called "planetary nebulas."

The nearest planetary nebula to ourselves is an object known as NGC 7293, which is about 85 light-years away. The diameter of the ring of gas is about a third of a light-year. These shells of gas are expanding, of course, growing thinner and thinner and eventually melt away into space generally.

Someday, after our Sun expands into a red giant, it will shrink and become a planetary nebula with a white dwarf at its center. (All planetary nebulae seem to have a white dwarf at their centers as nearly as we can tell.) The shell of gas will eventually disappear and what will be left will be just the white dwarf that was once our Sun, with at least 95 percent of its present mass, for not more than 5 percent would have been carried away by the gaseous shell.

Sirius B was undoubtedly a planetary nebula when it first formed, but that was so long ago that the gaseous halo has long since disappeared.

The fact that a red giant loses mass, either by spilling it over to a companion star as Sirius B did or by leaving behind a gaseous halo as we see in planetary nebulae (or both), is a vital point in the evolution of stars.

Such a conclusion is supported by the theoretical work of the Indian - American astronomer Subrahmanyan Chandrasekhar (1910—   ). Even in a white dwarf, there must be a balance between the gravitational force compressing the star and the temperature effect forcing its substance outward. The gravitational forces involved in a white dwarf are many times more intense than those of an ordinary star, and the central temperature must rise to balance it. The greater the mass of a white dwarf, the stronger the gravitational force compressing it and the smaller and denser it must be. At some critical point ("Chandrasekhar's limit"), no temperature would suffice to keep the white dwarf's structure from contracting to some ultimate limit. In 1931, Chandrasekhar showed that this critical limit was equal to about 1.4 times the mass of the Sun. This suggestion is supported by the fact that of those white dwarfs whose mass has been determined all are well below Chandrasekhar's limit.

The more massive the star, the more drastic the effect of compression and the more tremendous the explosions that result. Such explosions blow off large fractions of the mass and may reduce what is left to a value below Chandrasekhar's limit, allowing it to shrink to a white dwarf. As we shall see later in the book, this is not usually what happens. While stars of moderate mass do shrink, more or less quietly to white dwarfs, stars large enough to explode shrink more catastrophically to objects even more extreme in characteristics than the white dwarfs are. We will take up these extreme objects later in the book and, for the while, speak of all them together, if we have to, as "dwarf stars."

But exactly what sparks off the final collapse of a massive star? It happens so suddenly. At least two mechanisms for this sudden catastrophe have been offered and both may be valid, each in its own type of star.

One possibility is that the temperature of the iron-choked, innermost core eventually rises so high that, at a certain critical point, the iron atom is knocked apart by extraordinarily intense radiation into helium fragments. This is not a process that yields energy; it *absorbs* energy, all the energy contained in the radiation that sparked it. This happens to a certain extent at a wide range of temperatures, but at the critical point, it happens at such a rate that the energy absorbed can no longer be replaced by the last dregs of nuclear processes going on in the core. The temperature in the core drops slightly.

As the temperature drops, the star's gravitational compressive force, no longer completely countered, takes over. The star shrinks, and the core heats up at the expense of gravitational energy. The energy thus supplied keeps the iron-to-helium process going, and the star must shrink further at an ever-increasing rate In short, all the energy given off by the core over the space of some millions of years of helium fusion to iron must now be paid back in a matter of hours (!) from the only source of energy left— the gravitational field.

A second alternative was suggested in 1961 by the Chinese-American astronomer Hong-Yee Chiu (1932-   ). To explain this second alternative, some background must be introduced.

The ordinary radiation produced by a star's core is easily absorbed by matter. A fragment of radiation once released is quickly absorbed, released once more, absorbed again, and so on over and over. The radiation is passed from hand to hand, so to speak, in any direction at random, and it makes its way out toward the star's surface only very gradually. It is estimated that it takes an average of a million years for a given bit of energy to move from the Sun's core, where it is formed, to the Sun's surface where it is radiated away. This makes the substance of the Sun an excellent insulator, so that its center can be at 15,000,000° C., while its surface, less than half a million miles away, is only at 6000° C.

In addition to ordinary radiation, the nuclear reactions in the cores of stars produce extremely tiny particles called "neutrinos." Neutrinos travel at the speed of light, as ordinary radiation does, but there is this important distinction; neutrinos are only very rarely absorbed by matter. Any neutrinos formed in the Sun's center travel outward in all directions without being affected in any way by the matter making up the Sun. In three seconds, they reach the Sun's surface and then flash away into outer space, carrying some of the Sun's energy with them.

At the temperature of the Sun's interior, the number of neutrinos formed is minute in comparison to the ordinary radiation emitted. If the neutrinos are considered as a kind of energy-leak, so to speak, they form a virtually imperceptible one in the Sun.

As a star ages, however, the temperature of its center increases and the rate of neutrino-formation increases more and more rapidly. The neutrino energy-leak becomes more serious. When a critical temperature of 6,000,000,000° C. is reached (according to Chiu's calculations), the leak becomes so great that what nuclear reactions are carried on in the core are no longer sufficient to supply the energy required to keep the star from collapsing.

Whether the loss of energy at the core is caused by the sudden initiation of an iron-to-helium process or of an overwhelming neutrino energy-leak, the result, in either case, is the sudden catastrophic collapse of the star. In the process of collapse, the star's outer layers compress. These outer layers, however, still contain nuclear fuel—even hydrogen at the very surface. This hydrogen would not undergo fusion at the surface temperature of even the hottest stars, but with the added heat of compression, all the fuel remaining in the star is ignited. In a short space of time, energy that ordinarily might suffice for hundreds of thousands of years is radiated away.

If supernovas are the marks of the conversion of massive stars to dwarf stars, then surely there should be a dwarf star at the center of the Crab Nebula. As a matter of fact, a tiny, bluish star —as hot as one would expect a freshly formed dwarf star to be— does exist there. It was first detected by Walter Baade in 1942. For a quarter of a century it was considered a white dwarf, and then (as we shall see later in the book) it gained an even more exotic identity.

If there are indeed three supernova explosions per thousand years in a given galaxy, as Zwicky estimates, then in the 15 eons that our Galaxy has existed, there have been 45,000,000 supernovas formed. Add to these the dwarf stars that have formed more or less quietly through the planetary nebula stage and it isn't surprising that there are a considerable number of dwarf stars in the Galaxy. It may be that one out of a thousand stars have left the main sequence, expanded and then condensed to dwarfs.

Second-Generation Stars

When a large star explodes as a supernova, a quantity of matter pushes outward into space and, eventually, distributes itself among the thin gas already present there.

The exploded stellar matter is itself rich in all the elements up to iron since all these were present in the star at the time of explosion. Indeed, the exploded matter must also contain the elements beyond iron. These elements cannot be formed without an input of energy, and in an ordinary star they would not be formed. In a supernova, however, energy is produced at such an enormous rate that some is very likely to be put to work, so to speak, building up atoms more complex than those of iron. Atoms all the way to uranium, the most complex atom of those occurring in quantity in the Earth's crust, must be built up.

It is even very likely that elements more complex than uranium are formed. Such "trans-uranium elements" do not exist naturally on Earth because they are quite unstable. To be sure, uranium is itself unstable and is constantly breaking down, in stages, to lead. The uranium breakdown is so slow, however, that even after 4.5 eons (nearly the lifetime of the Earth) half the original supply of uranium remains intact. The breakdown of the elements beyond uranium is much more rapid, and if any of these elements existed at the time of the Earth's formation, they are all gone now.

In supernovae, however, there is at least one piece of evidence in favor of an at least transitory appearance of the element "californium" (six places beyond uranium in the usual listing of elements). Many supernovae seem to lose brightness with what seems a half-life of fifty-five days. It happens that one variety of californium, "californium-254," breaks down at just that rate. There may possibly be a connection here.

In any case, long after a supernova has flashed to its destruction and the resulting white dwarf has moved on in its journey around the galactic center, the gaseous shell of the explosion will linger behind through friction with the thin gas already present in interstellar space and serve to contaminate that gas in the region where the explosion had once taken place. A region of space which originally contained a thin gas that was exclusively hydrogen may now contain a small portion of helium and even smaller portions of more complex atoms. Indeed, if there are three supernovas per galaxy per thousand years that alone might be sufficient to account for one-third the interstellar gas.

What now, if a star is formed out of the matter of such contaminated space? It would still be mostly hydrogen; it would still have a long lifetime of many eons during which it could radiate, with hydrogen fusion as its energy-source (provided it were not too massive in the first place), but it would contain larger helpings of complex atoms than one might expect.

This seems to be the case with our Sun (and thus we answer

the problem raised on page 125). The Sun seems to be a "second-generation star," one that formed in an area where once a previously existing star had died explosively.

If the Sun had originally been 100 percent hydrogen, it would have taken 20 eons of time for it to decline to its present 81 percent of hydrogen. If it were only 87 percent hydrogen to begin with, it would have taken only 5 or 6 eons to reach its present state, and this may actually be what happened therefore.

The planets forming on the outskirts of the slowly agglomerating gas cloud that formed the Solar system would also end by being made up of material containing a considerable admixture of more complicated atoms. In the case of the Earth, which was too small and hot at the beginning to hold on to hydrogen and helium, the more complicated atoms make up almost the whole of the structure. The Earth has an inner core, making up fully one-third of its mass, which is almost entirely liquid iron. This seems to be an indication of the quantities of iron spewed into space by the exploding supernova with its own iron-choked innermost core.

Indeed, Fred Hoyle suggested some years ago that the Sun was once, like Sirius, part of a double star system and that the partner flashed into a supernova. In this way, Hoyle tried to explain the constitution of the planets and at the same time account for the fact that they possess so much angular momentum, since they would still possess much of the supply of the original exploded star. However, if we accept this hypothesis we have to solve the problem of the whereabouts of the white dwarf that ought to have been formed in the course of the supernova. Hoyle's suggestion is novel and interesting, but it is not considered at all likely by astronomers generally.

CHAPTER **12**

Galactic Evolution

The Question of Eternity

As you see, then, our consideration of the problems of stellar evolution has ended by suggesting that while the Sun might be only 5 eons old, there must be a Galactic history before that, for the Sun is built on the ruins of a still older star. Indeed, careful consideration of the structure of certain globular clusters, of the quantities of hydrogen they have consumed as compared with that which remains to be consumed, gives cluster lifetimes of up to 25 eons.

But is even that a necessary maximum? Is there any need to postulate any definite age for the Universe at all? A star might have a definite age, but there could have been stars, living and dying, before the present star in endless succession. One might consider the analogy of the human race which has existed for a period of time much longer than the lifetime of any single individual.

In a way, of course, we might argue that the energy of the Universe (including matter, as one form of energy) has always existed and always will exist since, as far as we know, it is impossible to create energy out of nothing or destroy it into nothing. This implies, we can conclude, that the substance of the Universe—and therefore the Universe itself—is eternal.

173

That, however, is not what we really mean. We are concerned with more than the mere substance of the Universe. The question here is whether that substance has always taken and will always take the form of the kind of Universe we know—one with stars and planets, and capable of playing host to living things like ourselves, or whether this Universe-we-know had a definite beginning and will have a definite end.

As far as we can tell now, the Universe-we-know (which is what I shall mean when I speak of the "Universe" henceforth) exists in this fashion on the basis of an energy output developed by hydrogen fusion. Before hydrogen fusion began, the Universe would have to be pictured (on the basis of the discussion in the previous chapters) as no more than a vast mass of whirling gas, perhaps dimly red-hot in spots. Once all the hydrogen has been fused, the Universe would have to consist of nothing but white dwarfs which have progressed for varying periods of time down the road to final darkness and extinction. It is only during the period when the process of hydrogen fusion is progressing massively that we have the Universe-we-know.

If we ask, then, whether the Universe might not be eternal, we are, essentially, asking whether hydrogen fusion might not go on forever.

Now we can see why, on page 95, I raised the question as to whether the Universe were infinite and then left the matter where it was while I veered off to take up the question of the age of the various heavenly bodies. For now I am raising the similar question as to whether the Universe is eternal, and it turns out that the two questions are closely related in some ways.

For instance, if the quantity of hydrogen in the Universe is infinite then it could, clearly, continue to fuse to form helium forever, provided the rate of fusion was finite. In other words, an infinite Universe ought to be an eternal Universe as well. On the other hand, a finite Universe might inevitably be finite in time as well as in space.

For a finite Universe to be eternal despite its finiteness there must be some process that can reverse the fusion of hydrogen, restore the hydrogen and make it available for fusion once more. Furthermore, this must not be done at the irrevocable expense of other energy sources, such as gravitational fields.

It must be stated that at first glance this does not seem possible. To be sure, according to firmly accepted scientific doctrine, energy is conserved and can never be destroyed. (This is sometimes called "the first law of thermodynamics.") However, although energy is always with us in constant quantities, it is not always available for conversion into useful work, and it is this availability for conversion into work that is a fundamental requirement for the Universe-we-know. Indeed, a generalization, which is known as "the second law of thermodynamics" and which seems to be as valid, universal, and important as the first law, tells us that the amount of energy available for conversion into work decreases constantly.

This means, to cite a very simple case, that water that has flowed downhill cannot flow uphill again by itself; it must be pumped uphill at the expense of energy from some other source. Any system that has run down, whether it has done work in the process or not, can be restored to its original state and allowed to run down again—but only at the expense of outside energy. Furthermore, it always takes *more* energy to restore a rundown system than the energy you would get by allowing it to run down again, so that there is always a net loss in the process of restoration. The second law is, as far as we know, inviolable.

We could conclude, then, that, little by little, more and more of the energy of the Universe will be tied up, irrevocably and irreversibly, into white dwarf remnants, and that will be the end of the Universe from our standpoint. Working backward, we might also conclude that although there were stars before the Sun, there must have been a time, say 25 eons ago, when *all* the energy of the Universe was in the form of thin wisps of swirling hydrogen, and that would be the beginning of the Universe from our standpoint.

By this line of argument the Universe might have a lifetime of perhaps 1000 eons; and of this enormous (but finite) length of time, a fortieth has already passed.

Yet this conclusion rests on the assumption that the first and second laws of thermodynamics are really valid everywhere in the Universe and not only in that small portion of it which we have been able to inspect, and that they are valid under all possible circumstances and not only those we have been able to witness.

Keeping in mind that the assumption of the validity of the laws of thermodynamics may be questioned, let us continue to try to find out whether the Universe is infinite and eternal, or finite and time-bound. To do so, let us turn from individual stars to the galaxies.

Classes of Galaxies

Notions as to the evolution of stars were first developed by a consideration of the different properties of different kinds of stars. The different properties were derived almost entirely from the spectra since the stars themselves were too small to show structural detail.

Notions as to the evolution of galaxies can also be developed by a consideration of the different properties of different kinds of galaxies. Here, though, there is a difference; galaxies are much larger than stars. Even though galaxies are far more distant from us than are the individual stars of our own neighborhood of space, these distant galaxies appear in our telescopes as more than points. They are patches of light which have particular shapes, and several thousand of the nearer among them even show considerable detail.

Hubble noted the three main classes of galaxies: spiral, ellipsoidal, and irregular (see page 94). He was able to do better than this, however, and in 1925 published a detailed classification that has been used ever since. The ellipsoidal galaxies, for instance, which lack spiral arms and look like very distant and very large globular clusters, differ among themselves in the degree of flattening. Some are virtually spheres (and might be called "spheroidal galaxies"), while some are rather flattened, some quite flattened, and so on. Hubble symbolized all the ellipsoidal galaxies as E and distinguished the degree of flattening by numbers. E0 represented the spheroidal galaxies, and E1 through E7 represented increasing degrees of flattening. An E7 galaxy is quite flattened, with its ends (if seen edgewise) sticking out point-fashion almost as though it were on the verge of possessing spiral arms.

As for the spiral galaxies, Hubble recognized two kinds. There was first the ordinary spiral in which the arms are directly connected with and wrapped about the ellipsoidal nucleus as in the case of the Andromeda galaxy. Then there were galaxies in which a straight bar of stars seems to extend outward from the nucleus on either side. From either end of this bar, there extend spiral arms. These are the "barred spiral galaxies," which seem to make up about 30 percent of all the spirals.

Hubble symbolized the ordinary spirals as S, and the barred spirals as SB. He then differentiated among the spiral galaxies of both classes according to how tightly or loosely the arms were wrapped about the nucleus, using lower case letters for the purpose: a for the most tightly wrapped arms, then b and c for looser structures. A spiral galaxy could be referred to as an Sa, Sb, or Sc; or, if barred, SBa, SBb, or SBc.

The Andromeda galaxy is classified Sb. Our own Galaxy has usually been considered similar to the Andromeda galaxy and has therefore been classified as Sb also. There may be a change here though. On the basis of measurements made in 1965 of the brightness of stars near our Galactic nucleus, it was suggested that the Galactic nucleus is smaller than had previously been thought. It may be only 6500 light-years across, half that of the nucleus of the Andromeda galaxy. If the nucleus of our Galaxy is indeed smaller than had been thought, and the spiral arms correspondingly more prominent and widely spaced, then our Galaxy would resemble the Whirlpool galaxy rather than Andromeda. Our Galaxy, like the Whirlpool, would then belong to the Sc classification.

Hubble arranged all these forms into a progressive order as follows:

$$E0 \rightarrow E1 \rightarrow E2 \rightarrow E3 \rightarrow E4 \rightarrow E5 \rightarrow E6 \rightarrow E7 \rightarrow S0 \quad \begin{array}{l} \nearrow Sa \rightarrow Sb \rightarrow Sc \\ \searrow SBa \rightarrow SBb \rightarrow SBc \end{array}$$

where S0 represents a hypothetical form with characteristics intermediate between ellipsoidal, spiral, and barred spiral.

Hubble did not specifically state that this represented, in his opinion, an evolutionary change, but apparently he felt it did. Certainly the gradual progression of change from spheroidal to loose spiral made an evolutionary hypothesis very tempting, and for a decade or so after 1925, the views on galactic evolution resembled a kind of Laplacian nebular hypothesis on a much vaster scale.

Imagine, to begin with, a quantity of gas not merely large enough to form a star, but large enough to form a hundred billion stars. Within this tremendous amount of gas, which we might call a protogalaxy, processes went on which led to its condensation into billions of stars. The protogalaxy would begin with a certain supply of angular momentum and as it condensed, it flattened. It would begin as a spheroidal galaxy and gradually flatten, passing from E0 to E7.

As the galaxy as a whole continued to contract and grow more compact, the rotation would become more rapid and so would the flattening progress until, just as in Laplace's planetary-system nebula, fragments would be given off at the equator. Spiral arms would form, with or without forming a bar (and neither Hubble nor anyone else has ever explained why the bar should exist at all). Furthermore, as time went on and the galaxy continued to contract and increase its rotation, the arms would continue to move away from the nucleus, so that both spirals and barred spirals would pass through the a, b, and c stages. The irregular galaxies might represent the final stage of this evolutionary scheme.

Since elliptical galaxies seem to be particularly large and irregular galaxies are usually small, we would have to suppose that the larger a galaxy the more slowly it passed through such a development. The very large galaxies would linger in the elliptical stage, while the very small galaxies would hasten through to the final irregular stage.

If this scheme is correct, then the Andromeda galaxy and, even more so, our own Galaxy, are relatively old, and far down the line of evolutionary development.

Stellar Populations

By the 1940's, however, new sets of views were arising. In 1942, Baade had an unusual opportunity. The city of Los Angeles was blacked out because of World War II, and it was possible to get a clearer look at the Andromeda galaxy than ever before, making use of the 100-inch Mt. Wilson telescope.

Until then, only the stars in the spiral arms had been made out individually by Hubble and by those who followed him. Now Baade was able to see and photograph stars in the Andromeda nucleus.

An important difference showed up. The brightest stars in the spiral arms were giant blue-white stars, very large and hot, something like the brightest stars in our own neighborhood of our own Galaxy. The brightest stars of the Andromeda nucleus were, however, reddish stars; there were no blue-whites at all.

This seemed to fit in with general spectral information. The spectrum of the Andromeda nucleus, and of those other galactic nuclei that could be examined, as well as the spectra of ellipsoidal galaxies generally, tended to be of spectral class K. The average starlight of those regions tended to rise from surfaces dimmer and cooler than that of the Sun. The general spectrum of the spiral arms of Andromeda and of other galaxies tended to be of spectral class F. The average starlight of those regions tended to rise from surfaces brighter and hotter than that of the Sun.

It was as though regions where stars were concentrated closely together, as in globular clusters, galactic nuclei, and ellipsoidal galaxies generally, were largely of one type, which Baade called "Population II." On the other hand, stars that were more loosely spread out, as notably in the spiral arms of galaxies, were largely of another type, which he called "Population I."

The Population II stars seemed to be distributed in a kind of spherical halo about the Galactic center, while the Population I stars seemed to be distributed in a kind of hollow disc along the central plane of the Galaxy. They might be referred to as "halo stars" and the "disc stars," respectively.

On the whole, Population II stars tend to be sedate and uniform, moderate to small in size and occupying regions of space that are relatively free of dust and gas. Population I stars, on the other hand, tend to include numbers of rather spectacular members and to show a wide variety, including large stars, more brilliant and more fiercely hot than anything among the Population II stars. Furthermore, Population I stars occupy regions of space that are relatively rich in dust and gas.

This division into two populations (a division that has since been made more complicated with Population I and II both broken down into several subclasses) raised new questions concerning galactic evolution. Gathering knowledge concerning the nature of nuclear reactions within stellar cores made it seem that large stars were shorter lived than small stars so that Population I stars had to be viewed as rather evanescent. Closer analysis showed that the spiral arms themselves had to be short-lived, and no completely satisfactory explanation has yet been advanced to account for the fact that so many galaxies seem to sport spiral arms if the latter are as short-lived as they ought to be.

In any case, the feeling grew that spiral arms had to be temporary phenomena and that galaxies tended to lose them and become elliptical, rather than the reverse. Hubble's evolutionary scheme was there turned upside down.

Suppose that a protogalaxy formed stars in irregular fashion at first, so that the earliest stage was that of an irregular galaxy. At the center of the protogalaxy, where the dust was most concentrated, stars would form most rapidly and numerously. There would be a relatively small supply of dust and gas per star, and the stars of the center would be on the small side and rather uniform in properties. And when they were formed, the dust and gas would be almost gone. Such stars would be of the Population II type, very rich in hydrogen, very low in the more complex atoms.

On the outskirts of the protogalaxy, however, the dust and gas would tend to be less evenly distributed. Perhaps the general rotation of the protogalaxy whipped the dust and gas into shreds, so that stars tended to form in lines and regions that turned out to be luminous spiral arms. Because of the uneven distribution of dust and gas, stars would form numerously here and sparsely there; some would have an unusually small share of material to begin with, some an unusually large share. There would thus be a great variety in their masses, and the large members of this Population I group would be bright and hot.

In the spiral arms, moreover, much dust and gas would be left over, having been spread out too thinly in their far-from-the-center location to serve as nuclei for stars. Over eons of time they slowly condensed until such time as they could finally develop into stars. Furthermore, the bright, hot Population I stars would in the course of their speedy evolution reach the death-stage and explode as supernovae. The dust and gas in the spiral arms would come to be enriched with helium and more complex atoms, so that Population I stars which developed later would be second-generation stars, relatively poor in hydrogen and rich in more complex atoms.

Thus, the first stage of galactic evolution after the original irregular galaxy would be a spiral galaxy with the arms loosely wrapped.

But the spiral arms are relatively short-lived. The dust and gas are consumed in star formation, the brighter stars die out, and the situation in the arms progresses steadily from a Population I situation to a Population II situation. Furthermore, as a galaxy revolves, the spiral arms trail and tend to wind up, so to speak, coming to hug the nucleus more and more closely, progressing from Sc to Sb to Sa.

Eventually, the spiral arms melt into the nucleus, their Population I character completely gone, and a flattened elliptical galaxy is formed.

With time, the stars of such a galaxy, interacting gravitationally with one another, spread their motions more and more evenly through the whole, so that there is less and less flattening until a spheroidal galaxy is formed.

Looked at in this fashion, it would seem that the Andromeda galaxy and our own are not in the later stages of evolution, but in the earlier, and are, in fact, rather young galaxies.

The reverse-direction scheme of galactic evolution has some observations in its favor. It postulates a steady diminution of dust and gas in galaxies (which seems reasonable); irregular galaxies such as the Magellanic Clouds are 20 to 50 percent gas and dust. They seem to be more debris-filled than our own definitely spiral Galaxy, while spiral galaxies are in turn dustier than the elliptical galaxies.

To account for the large size of so many spheroidal galaxies, it might be suggested that large galaxies (like large stars) evolve more quickly rather than more slowly. The largest, under the influence of a particularly strong gravitational field, would more quickly have condensed into stars, more quickly formed spiral arms in more dust-filled outer regions, more quickly dragged the arms inward and rounded itself out to form the giant spheroidal galaxies we now have. Smaller galaxies would linger in the spiral form, and still smaller galaxies would stretch out the early irregular phase of the evolutionary process.

Then, again, the possibility is advanced that evolution in the ordinary sense—with a galaxy of one kind turning into a galaxy of another kind—may not take place at all. Once a galaxy has taken on its star-filled shape, it is suggested, that shape is set, and the difference between galaxies depends entirely on differences in the original protogalaxies, chiefly in the matter of the quantity of angular momentum present.

Suppose a protogalaxy happens to have very little angular momentum. It rotates slowly and does not flatten much, if at all. Little matter would be lost through the centrifugal effect, so that such a protogalaxy would retain a maximum size. Stars would form and make up a vast spheroidal galaxy.

If the protogalaxy happened to have a larger supply of angular momentum, it would turn more rapidly, flatten somewhat, leak away some matter at the equator, and end up as a somewhat smaller, somewhat flatter ellipsoidal galaxy.

Changes would end once stars formed and, depending on the quantity of angular momentum, different amounts of flattening would take place before star-formation froze the shape. In general, the more flattened the galaxy, the smaller.

If the protogalaxy had a particularly large amount of angular momentum, a new factor would enter. Stars would begin

Formation of star populations

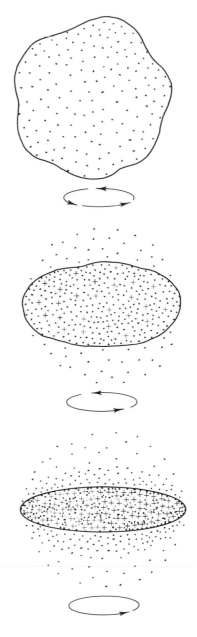

+ = population II
• = population I

forming but not at a rate fast enough to wipe out the dust and gas before the rapid rotation succeeded in flattening the proto-galaxy considerably further. In short, the gas and dust would flatten into a disc, leaving behind a spherical or ellipsoidal halo of stars. The stars left behind would be the Population II stars.

The disc of gas would be thrown outward into the extremi-ties and concentrate there, and within it would be formed the Population I stars. Thus the formation of spiral galaxies and of Population I stars would be entirely a matter of the rate of rota-tion of the original protogalaxy.

Whatever scheme of galactic evolution is considered—whether one progresses from protogalaxy to spheroidal to spiral, from protogalaxy to spiral to spheroidal, or from protogalaxy to spiral and spheroidal independently—one is left uncertain as to the age of the galaxies generally and, therefore, as to the age of the Universe.

Did the entire Universe form from a system of protogalaxies that all (about 15 eons ago) began to develop into galaxies? If so, then, the Universe as a whole is 15 eons old.

Or did some parts of the Universe begin galactic develop-ment at some times, and other parts at other times? Is that por-tion of the Universe with which we are most familiar 15 eons old, while other portions are older? Are there some vast regions, unde-tectable to us, that are still in the protogalaxy stage and that, in effect, have not been born yet?

Such questions do not get a ready answer from considera-tions of galactic evolution alone. We must approach the problem, therefore, from another direction.

13

The Receding Galaxies

The Galactic Red Shift

A key step in the study of the galaxies was taken in 1912 when the galaxies were not yet recognized as such but were considered as spiral nebulae within our own system of stars. In that year, the American astronomer Vesto Melvin Slipher (1875-1969) measured the radial velocity of what was then called the Andromeda Nebula and found it to be approaching us (its spectral lines showed a violet shift) at a speed of 200 kilometers per second.

This was interesting but not unusual. Radial velocity of over a hundred kilometers per second was high but not distressingly so. (In fact, we know nowadays that part of the velocity is not to be attributed to a true approach of the Andromeda. The rotation of our Galaxy happens to be carrying the Solar system toward Andromeda, just as millions of years from now it will be carrying us away from it. If the effect of this rotation is allowed for and the motion of Andromeda is measured relative to the center of the Galaxy, it is still found to be approaching us, but only at a velocity of about 50 kilometers per second.)

By 1917, things seemed somewhat more puzzling. Slipher had gone on to measure the radial velocity of a total of fifteen spiral nebulae. Offhand, on the basis of sheer chance, one might have expected half of them to be approaching, half receding. Instead, two were approaching and thirteen were receding. Although this is not the most likely situation, it is still not impossible. (If you toss fifteen coins, you might expect a roughly equal number of heads and tails, but chance is chance—you might get two heads and thirteen tails.)

More disturbing was the fact that the radial velocities were as great as they were. Those nebulae, which were receding, were doing so at an average velocity of 640 kilometers per second. While 200 kilometers per second could be swallowed, 640 kilometers per second was quite difficult to accept. It was much greater than the radial velocities of the ordinary stars of the galaxies.

The more Slipher measured the radial velocities of the spiral nebulae, the more extreme the situation grew. Each new measurement he made showed a recession, never an approach, and with increasingly astonishing velocities.

When Hubble demonstrated, in the mid-1920's, that the objects Slipher was observing were really galaxies that were located far outside our own Milky Way, this, in a way, eased the problem. That one particular class of objects making up our Galaxy should have such large radial velocities of recession, while all other objects had small velocities that were often velocities of approach, made no sense at all. If, on the other hand, the objects showing the unusual velocities were also at unusual distances, things were better. If *two* properties of an object are unusual, there may be a connection and each can help explain the other.

The work on the radial velocities of the galaxies was now taken up by another American astronomer, Milton La Salle Humason (1891–1972). With the greatest care he began taking photographic exposures of days at a time so that fainter and fainter galaxies could record their spectra. He discovered velocities of recession among the faint galaxies that made earlier determinations seem most conservative. In 1928, he tested the radial velocity of a galaxy called NGC 7619 and got a value of 3800 kilometers per second, and by 1936 he was clocking velocities of 40,000 kilometers per second, better than one-eighth the speed of light— and they were always recessions.

Such velocities were so great that astronomers fell to questioning the nature of the red shift (and occasionally the matter is still brought into question). Must a red shift necessarily imply that the source of light is receding? Or could there be some alternate explanation that would make it unnecessary for us to accept such huge velocities?

Could it be, for instance, that the light of the distant galaxies is simply reddened by incredibly long passages through the thin gas of intergalactic space. Undoubtedly, but such reddening of light is not at all the same thing as a red shift. Short-wave light would be removed from the spectrum by such scattering and the color of the nebula as a whole would be reddened, but the spectral lines would not be shifted at all. And it is to the spectral lines particularly, not to the color generally, that the term redshift is applied.

Another suggestion sometimes made is that the light, on its unprecedentedly long journey from the galaxies, somehow loses energy in transit. When light loses energy, this loss is displayed in terms of an increase in wavelength so that one would expect, under such conditions to obtain a true red shift. We would thus be victimized into thinking the galaxies were receding at unprecedented velocities simply because we were studying what might be called "tired light."

This explanation does not explain, however, because no one has yet offered a mechanism that would account for the postulated loss in energy. There is no known reason why light should lose energy simply because it was traveling through a vacuum for a long time. Furthermore, if it were indeed losing energy in this fashion, no one could offer a reasonable explanation as to what became of that energy. Then, too, if light grew tired over long distances, it ought to grow very slightly tired over shorter distances. Judging by the red shift related to the galaxies, the red shift that would be produced by objects closer than the galaxies ought to be detectable, but it is not.

In short, the red shift can, so far, only be explained by supposing the galaxies to be receding from us. No alternate explanation can square with all the facts and make sense. Until an alternate explanation does, astronomers will have no choice but to continue to accept the precipitous recession of the galaxies as a fact.

Hubble, working along with Humason, was naturally interested in the recession of the galaxies. He had been painstakingly making estimates of the distance of the galaxies in several ways. He had made use of Cepheids in the closest (see page 92). For those too far away to reveal any Cepheids, he made use of any stars he could see by assuming these were supergiants that were, say, as bright as S Doradus. If the galaxies were too far away for any stars to be made out at all, then he made the assumption that, on the whole, galaxies were approximately equal in total luminosity and that the dimmer a galaxy was, the more distant it was. Naturally, one applied the inverse square law. If one galaxy was one-fourth as bright as another, it was twice as far; if it was one-ninth as bright, it was three times as far, and so on.

Using such criteria of distance, Hubble, in 1929, made use of the velocity determinations of Slipher and Humason to show that, on the whole, the velocity of recession of the galaxies increased proportionately with the distance of those galaxies from us. If one galaxy was twice as far from us as another, it receded from us at twice the velocity; if it was three times as far from us as another, it receded from us at three times the velocity, and so on. This is "Hubble's law."

The most astonishing feature of Hubble's law that the velocity of recession of a galaxy is proportional to its distance from us can best be expressed by the simple question, "Why *us*?"

What magic is there about ourselves that causes a galaxy to hasten from us, so to speak? And how does the galaxy "know" how far it is from us and guide its steps accordingly?

Fortunately, the same explanation that accounts for the relationship between speed of recession and distance in the first place, also accounts at once why it should be distance from *us*.

The explanation was suggested by the new view of the Universe presented by Einstein.

Relativity

The new view was contained in Einstein's general theory of relativity, first put forth in 1915. In it, Einstein worked out a set of "field equations" which described the overall properties of the Universe. In order to do this, he assumed that although the Universe showed condensations of matter here and there (planets, stars, galaxies), it could be treated with approximate accuracy if it were regarded as though it were uniformly filled with matter; that is, as though the matter actually in the Universe were spread out evenly. (This is analogous to the manner in which, despite the fact that the surface of the Earth is manifestly bumpy and irregular, ancient man made the assumption that all these irregularities were not really of importance and that, on the whole they could be looked on as though they were spread out evenly so that the world could be considered flat. We do the same sort of thing today except that we consider the world to be a sphere.)

Einstein further assumed that the properties of the Universe were, generally speaking, the same everywhere. On the basis of this assumption, the possible geometries of the Universe were strictly limited. To see why, let us use the surface of the Earth as an analogy.

Wherever we are on Earth, we feel essentially the same. The directions up and down are the same; the pull of gravity is about the same; the horizon is always at the same distance, and is equally so in all directions (assuming that we disregard local irregularities in the surface and consider all matter spread out evenly).

There are three kinds of surfaces the Earth could have that would yield identical properties of this sort everywhere. It could be planar (that is, flat); it could be spherical; or it could have a much less familiar shape called "pseudospherical." Early man assumed the surface to be flat because that was simplest, but observation eventually made it necessary to choose the spherical shape instead.[1]

The choice of the spherical surface of the Earth over the others has an important geometric consequence. The spherical surface is the only one of the three that is finite. A straight line on a flat surface, or the equivalent of a straight line on a pseudospherical surface, would go on forever. The equivalent of a straight line on a spherical surface, however, would close in on itself. That is, if you begin at some point on the Equator and walk east, you will eventually return to your starting point although you have never changed direction. You can walk on forever without coming to any "end of the Earth," but you will be repeating your path endlessly. The surface of a sphere is finite, but unbounded.

The same situation can be applied to the Universe as a whole, except that in the Universe we are dealing with a volume, rather than with a surface, and that makes the matter harder to visualize.

Still, consider a ray of light traveling through the Universe. To us it seems that a ray of light traveling through a perfect vacuum and encountering no interfering energy fields must travel in an absolutely straight line forever, receding from its source at a constant rate. This is equivalent to saying that the Universe

---

[1] Of course, the Earth is not *exactly* spherical, and therefore properties such as gravitational intensity at the surface do vary slightly from place to place.

has the properties one would describe by means of Euclidean geometry. We could call it a "flat Universe" even though it is a volume rather than a surface.

But is the Universe really Euclidean or is that an illusion born of the fact that we see so small a portion of it? A small portion of the Earth's surface looks flat to us, too, and only very delicate measurements tell us that it is actually gently curved in every direction.

But if the Universe is not Euclidean, what can it be? If we assume its general properties are the same everywhere, we have the same choices we had in connection with the Earth's surface, if we consider it in terms of a ray of light. There are two alternatives to the flatness, two varieties of a non-Euclidean Universe.

The ray of light can move in a grand circle, as though it were moving along the surface of a sphere. The geometry of the Universe would then correspond to a system first described by the German mathematician Georg Friedrich Bernhard Riemann (1826–1866) in 1854. A Riemannian Universe must not be looked on as a simple spherical Universe; it is more complicated. It is one in which three-dimensional space itself curves in every direction with a constant curvature. The Universe is a four-dimensional analogue of a sphere, a "hypersphere"—something that is very difficult to represent or imagine, accustomed as we are to thinking in three-dimensional terms.

The ray of light can also move as though it were following a pseudospherical surface in all directions. The geometry of the Universe would then correspond to a system first described by the Russian mathematician Nikolai Ivanovich Lobachevski[2] (1793–1856) in 1829.

The Riemannian Universe differs from the Euclidean and Lobachevskian Universes in being finite. A ray of light traveling through a Riemannian Universe curves back on itself. It can go on forever but only by endlessly repeating its path, like the Earth's equator. The Riemannian Universe is therefore finite but unbounded.

How can one choose among these three possibilities? If we could make a ray of light travel over a long enough distance, we could perhaps see whether it was actually straight, or whether it was deviating from a straight line in either a Riemannian or

[2]   The satirist Tom Lehrer has written a very funny song entitled "Lobachevski." Those who have heard and enjoyed the song may think I have made a mistake, but Lehrer made use of the name of a real—and great—mathematician.

Lobachevskian sense. The Universe deviates from the Euclidean so slightly (assuming it deviates at all) that a ray of light long enough for the purpose would be difficult to handle. Worse than that, we would be stymied in our search for straightness by the fact that our criterion for straightness is light itself.

If we have a long measuring rod and want to know if it is straight or not, we can hold it up endwise to our eyes and sight along it. If it is not straight, we see it dip below the line of sight or bulge above it or bend to one side, and even small deviations from straightness are quickly detected. But what we are doing in this case is to make the assumption that rays of light are traveling in an absolutely straight line. Our assumption of the straightness of light is so absolute that when light is reflected or refracted, nothing can convince our sense of sight that the straightness has been violated. We see ourselves *behind* the mirror; we see a stick bend sharply where it enters the water.

The decision among the possible Universe must therefore be made more indirectly. Einstein chose the Riemannian Universe and, by 1917, had worked out its consequences, trying to find some that would deviate measurably from similar consequences in a Euclidean or Lobachevskian Universe. (This can be considered as marking the beginning of the modern science of cosmology.) For instance, he argued that light would lose radiational energy in traveling against a gravitational field, and this Einstein shift he predicted was found in the light radiated by Sirius B (see page 160). He also predicted that light rays would bend on passing a massive object, so that stars would seem to shift position slightly if their light passed near the Sun. The positions of stars near the Sun were measured during a total eclipse in 1919 and compared with the positions at times when the Sun was nowhere near, and again Einstein was borne out. In all tests made of Einstein's general theory of relativity, the theory has been borne out. Not one observation going clearly against it has been made, and it is generally accepted among astronomers that the Universe as a whole follows a Riemannian geometry although one that deviates so slightly from the Euclidean that under ordinary circumstances Euclidean geometry is perfectly satisfactory.

Einstein further pictured a Riemannian Universe that was static, one that did not undergo an overall change. The individual components within it might move around, but the overall density of matter, if all were smoothed out evenly, would remain the same. Since, in Einstein's view, the curvature of the Universe (the extent to which it was Riemannian) depended on the density, a ray of light would travel in a perfect circle if uninterfered with.

In 1917, however, the Dutch astronomer Willem de Sitter (1872–1934), who had been among the first to accept relativity, offered another design of the Universe that would also fit Einstein's field equations. This was a Universe that was empty and constantly expanding. This meant that the curvature of space was becoming constantly less marked; the Universe was Riemannian but was perpetually approaching closer and closer to the Euclidean (which it would reach when expansion had grown infinite). A ray of light in de Sitter's model of the Universe would not travel in a circle but in a perpetually expanding spiral.

Furthermore, suppose one were (in imagination) to insert two particles in de Sitter's expanding Universe. These two particles would separate at once and would continue separating as the space between them expanded and continued to expand.

If a large number of particles were scattered through such a Universe, the general expansion of the Universe would increase the distance from any one of them to any other. If the distance between a given particle and its nearest neighbor was 1 light-year to begin with, that distance would be 2 light-years after a while, 3 light-years after another while, and so on.

Now suppose an observer were on one of those particles, looking at all the others. In one particular direction there would be a particle 1 light-year away, another one beyond it which would be a total of 2 light-years away, still another 3 light-years away, and so on. After a century, let us say, the distance between any two neighboring particles has increased to 2 light-years. By that time, then, the observer standing on his particle and looking in the same direction as before would see the nearest particle 2 light-years away, the next 4 light-years away, the next 6 light-years away, and so on.

If that is the case, then the nearest particle has moved outward from 1 light-year to 2 light-years and has receded at the rate of 1 light-year per century. The second particle has moved from 2 to 4 light-years and has receded at the rate of 2 light-years per century. The third particle has moved from 3 light-years to 6 and has receded at 3 light-years per century.

All the particles you see in a particular direction are receding from you and at a rate that is proportional to their distance. Furthermore this is true in no matter which direction you look. Does that make your particle special? Not at all. It would not matter which particle you were on. The effect would be the same on any one of them. Every independent particle in such a universe is receding *from every other particle* at a rate proportional to the distance between the two particles being considered.

De Sitter's expanding Universe seemed quite superior to Einstein's static Universe from a theoretical standpoint. This superiority was strengthened further in 1922, by a Russian astronomer, Alexander Friedman, who applied it to a non-empty Universe. And then, in 1930, Eddington was able to demonstrate that even if Einstein's static Universe could be assumed to exist it would be unstable, like a cone balanced on its point. If it began to expand ever so slightly for any reason, it would continue to expand forever; and, for that matter, if it began to contract, it would continue to contract indefinitely.

Clusters of Galaxies

Hubble's law seems to show the existence, then, in real-Universe terms, of de Sitter's theoretical model of an expanding Universe. The galaxies are receding from each other not because they are moving individually, but because all of space is expanding. As a result of this expansion, the velocity of recession of an individual galaxy is proportional to its distance from us. Moreover, our Galaxy seems the central point of the universal recession only because we happen to be observing the Universe from that point. Had we been in the Andromeda galaxy, or in any other galaxy, the same phenomenon would have been observed. The galaxy in which we were would seem the central galaxy in every case.

Of course, one might say that if the Universe were really expanding, then *every* galaxy without exception ought to be receding from every other galaxy, and that while this is almost true it is not completely true, and "almost true" wins no prizes. For instance, the Andromeda galaxy is not receding from us; it is approaching us. It is doing so quite slowly, to be sure, but it is approaching and surely that upsets the entire notion of an expanding Universe.

Not at all. In presenting a model of the Universe, one is forced to make simplifying assumptions or else the complexities of the model become too great to be analyzed or described. In the de Sitter model, for instance, it is assumed that the test particles introduced exert no force on each other, but accompany the expansive movement of the Universe without resistance.

But this is not so in reality. There are long-range forces in the Universe that are capable of making themselves felt over great distances as soon as matter is considered to exist in it. These long distance forces are of two kinds. There are electromagnetic fields and gravitational fields. The electromagnetic field gives rise to forces of two kinds, an attractive one and a repulsive one, and the two effects usually balance on a large scale. We can therefore ignore the electromagnetic field in the Universe generally.

Not so in the case of the gravitational field. That gives rise to only one kind of force, an attractive one. Any two objects in a Universe, even in an expanding Universe, will attract each other.[3] .The closer two objects are, the stronger the gravitational attraction between them and the more likely they are to cling together, so to speak, against the separating influence of the expanding Universe.

The expanding Universe does not, for instance, separate the components of the Solar system from each other, or separate the stars within a Galaxy from each other. It will not suffice to separate two or more galaxies which are close enough together to be caught in the grip of a sufficiently strong mutual gravitational field.

In short, the "independent particles" that recede from each other in an expanding Universe are, in our own real Universe, not necessarily individual galaxies, but individual groups or clusters of galaxies.

Such clusters of galaxies clearly exist. There are a number of cases of two or more galaxies that visibly interact, that are enclosed in common halos or are connected by luminous threads. The cause of the interaction may be gravitational or it may be electromagnetic, but the interaction is there in either case and it is quite logical to treat such galaxies as a unit, from the overall standpoint of the Universe.

Our own Galaxy has the two Magellanic Clouds as obvious satellite galaxies that are firmly in our grip (there is even evidence that thin wisps of gas stretch between them and us) and are not likely to be separated from us by the expanding Universe. In the same way, the Andromeda galaxy (M31) has two small galaxies (M32 and M33) which are satellites to it.

In fact, the Andromeda galaxy and our own Galaxy may be considered as the two giant members of a group made up of some thirty members, mostly small, altogether. This group is the

---

[3]   There are special cases where this is not so as, for instance, if one object is a hollow sphere and the second object is within the hollow. These special cases are not important, astronomically.

"Local Group," and no doubt it will maintain its identity, at least over some long period, against the expansive pull of the Universe.

The motions of the member galaxies of the Local Group, relative to each other, do not reflect the general expansion of the Universe, but, rather, local gravitational forces. It is for this reason that the Andromeda galaxy happens, in this particular eon of time, to be approaching us.

There are hundreds of clusters of galaxies visible in the sky. They are obviously clusters because of the close propinquity of the individual members in space and of the similar luminosity of the larger members. Some of the clusters are enormous. There is one cluster in the constellation Coma Berenices, about 120,000,-000 light-years distant, that is made up of about 10,000 individual galaxies.

Such clusters are very useful in estimating distances. If galaxies are studied individually, it is not entirely safe to suppose that the degree of dimness is determined only by distance. There are giant galaxies and dwarf galaxies just as there are giant stars and dwarf stars. The Andromeda galaxy and its two satellite galaxies are at the same distance from us, yet Andromeda is much

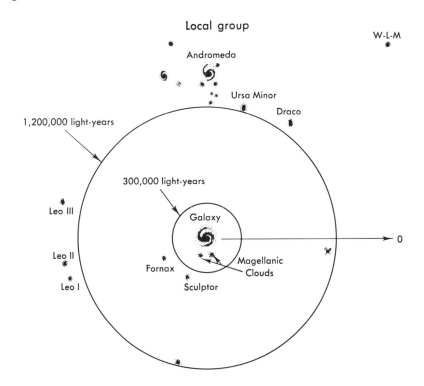

brighter than either of the other two, simply because it is a giant where the others are dwarfs.

Of course, one could argue that, over the long run, such differences in size average out and that, viewed as a whole and from a generous statistical standpoint, dimness can be equated with distance. This is true, but the general equation of dimness with distance suffers from irregularities as a result.

In clusters, however, one can assume that the brightest members are giant galaxies equivalent to the Andromeda or to our own and that they therefore have a total luminosity equivalent to an absolute magnitude of —19 or —20. There is still the chance of differences and irregularities from one cluster to the next, but these are much smaller, astronomers believe, than would be the case if one were dealing with individual galaxies.

Astronomers therefore feel more confident in judging the distance of clusters of galaxies from dimness alone, and then comparing that distance with the value of the red shift to see how Hubble's law holds up. (It does.) A cluster of galaxies in the constellation Virgo is estimated by this method to be over fifteen times as far away as the Andromeda galaxy, and other clusters are located at nearly a thousand times the distance of the Andromeda.

14

The Observable Universe

Olbers' Paradox Again

The scale of the Universe is seen to be so large that it dwarfs even the first consideration of the distances of nearer galaxies such as Andromeda. When the distance of the Andromeda galaxy was first determined and expressions such as "hundreds of thousands of light-years" came into vogue, that was considered mindstretching indeed. Within a decade or so, however, it became obvious that Andromeda was merely next door. It is even, as I have said, part of the Local Group, part of a system of which we are also part.

One has to ask again where it ends. Time and again, man has had to expand his vision to take in larger and larger groups. Small nonluminous objects group together about a star to form a planetary system. Stars group together to form a simple multiple-star system, or larger open clusters, or even larger globular clusters, or still larger galaxies. Galaxies group together to form clusters of galaxies. Can these group together still more extensively to form clusters of galaxies? De Vaucouleurs suggests that this may be so, that there are signs that a "supergalaxy" may exist in which the Local Group is but a small item. If his analysis is correct, then we are tens of millions of light-years from the center of such a supergalaxy—and beyond it in every direction would be other supergalaxies.

And would there not be clusters of supergalaxies and clusters of clusters of supergalaxies and so on? What would be the end? Would there have to be an end at all? May we not be facing an infinite Universe? Such a system of unending clusters of clusters is called a "hierarchy Universe" and was first suggested by a Swedish astronomer, Carl Wilhelm Ludwig Charlier (1862–1934).

To be sure, if we accept Einstein's theory of relativity then his Riemannian Universe must have a finite volume. Even if it is expanding, that finite volume, although constantly growing, remains finite.

And yet it is sometimes argued that even though the Universe may have a finite volume, it might be able to hold an infinite number of galaxies. If that is so, then the system of clusters of clusters of clusters of clusters of galaxies may continue onward into greater and greater complications without end.

But if we are going to consider the possibility of an infinite number of galaxies, do we not again run into Olbers' paradox (see page 44)? Will not the existence of an infinite number of galaxies in every direction supply the Earth with an infinite amount of light? And from the fact that the Earth does not receive an infinite amount of light must we not argue that the number of galaxies must be finite?

Supergalaxies

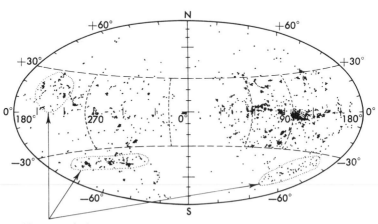

possible supergalaxies

We must admit a finite number of galaxies, indeed, *if* the Riemannian Universe were a static one, as Einstein had first proposed. In such a static Universe, the argument I will later advance in favor of the possibility of an infinite number of galaxies would not hold. In such a static Universe, both volume and the number of galaxies would be finite and Olbers' paradox would pose no problem.

But we seem to be living in an expanding Universe, and here the argument for an infinite number of galaxies *can* hold. What now? How do we get out from under Olbers' paradox.

In an expanding Universe, in which the galaxies recede steadily from each other, there is a new factor which must be taken into account, one which would not be present if the Universe were static and the galaxies maintained themselves at distances from each other that did not (on the average) change. That new factor is the red shift.

In an expanding Universe, the light from the galaxies is weakened and enfeebled by the red shift. The further the galaxy the stronger the red shift and the greater the weakening of the total radiational energy reaching us.

If we allow for the steady weakening of the radiation reaching us from greater and greater distances, then the total amount of radiation that impinges on the Earth can be shown to reach some finite value, and one that is not excessively large, even though the number of galaxies was infinitely large. Olbers' paradox would no longer stand in the way of an infinite Universe (in terms of numbers of galaxies) if that Universe were expanding.

This may sound impossible. It may seem to you that if every one of an infinite number of galaxies contributes a little radiation, then, no matter how little that might be, the sum total must be infinite. This is equivalent to saying that the sum of an infinite series of numbers, however small the individual numbers might be, must be infinite.

This may sound logical, but it is wrong just the same, as can be demonstrated very easily.

Consider the series of numbers: 1, 1/2, 1/4, 1/8, 1/16, 1/32, 1/64 . . . Each number is half the number before, and there are an infinite number of members of this series. No matter how far the series is extended it can always be extended farther by writing down half the last number written, and then one more by writing down half the additional number written, and so on indefinitely.

One might suppose, then, that the sum of such an infinite series must be infinite — but try it. The first number taken by itself is 1; the sum of the first two numbers is 1½; the sum of the first three is 1¾; the sum of the first four is 1⅞. If you continue onward, you will satisfy yourself soon enough that although the sum continually increases as more and more members of the series are taken, that sum can never reach 2. It approaches 2 more and more closely but never quite reaches it. The sum of the series $1 + 1/2 + 1/4 + 1/8 + 1/16 + 1/32 + 1/64 \ldots$ is said to *approach* 2. This is an example of a "converging series," a series with an infinite number of members but a finite sum.

In the original description of Olbers' paradox, I explained that in an infinite Universe of the type envisaged by Olbers each shell of space delivers an equal amount of light to Earth. If the light of one shell is considered 1, then the light of all shells is $1 + 1 + 1 + 1$, and so on forever. This is a "diverging series" and obviously has an infinite sum. It is that infinite sum that is the nub of the paradox.

If the red shift is taken into account, however, each shell (working outward) delivers less light than the one before. If a series is set up, it converges and yields a finite sum. In Einstein's Riemannian Universe, whether static or expanding, we can therefore forget about Olber's paradox. It presents no problem.

But now something else comes up. If we are going to consider the red shift and the consequent progressive weakening of the energy of radiation from more and more distant galaxies, we must ask ourselves not how large the Universe is, but how much of the Universe we can observe.

Man has taken it for granted that if he improves his instruments and refines his techniques, if he builds bigger telescopes and better spectroscopes and more delicate cameras, he will be able to see farther and farther out into space — but is this so? If the radiation from distant galaxies becomes feebler and feebler, is there a point at which the radiation becomes so feeble that no instrument, however close to perfection it may be, can detect it?

If this is so, then there is some limit, in principle, to the size of the Universe we can observe; there is some outermost boundary we cannot peer beyond. This boundary would delimit what we might term the "observable Universe."

To find out if there is such an inviolable limit and where it might be, let us return to Hubble's law.

Hubble's Constant

Hubble's law states that the velocity of recession of a galaxy is directly proportional to its distance from us. This means that if its distance from us is multiplied by some definite quantity (a quantity called "Hubble's constant"), we will get the velocity of recession.

Suppose we represent the distance as so many millions of light-years and call that D. We can represent the velocity of recession as so many miles per second and call that V. For the proportionality constant we can write k. Now we can express Hubble's law as follows:

$$V = kD$$

where kD is the mathematical way of saying that k must be multiplied by D.

Everything depends on the value of k, so let us rewrite the equation above, making use of simple algebraic techniques, and get:

$$k = V/D$$

which tells us that k must be equal to the velocity of recession of a galaxy (in miles per second) divided by its distance (in millions of light-years). If we have good reliable figures for both the distance and the velocity of recession of a single distant galaxy or group of galaxies, then we can solve for k. That value of k will, if Hubble is correct, hold for all galaxies.

Consider the Virgo cluster, for instance. The red shift of its components shows it to be receding from us at a velocity of 710 miles per second. By comparing the brightness of the brightest galaxies of the cluster with that of the Andromeda galaxy, the cluster turns out to be 16.5 times as far distant as the Andromeda. If the Andromeda galaxy is 800,000 light-years away (according to the period-luminosity relationship of the Cepheids, see page 57), the Virgo cluster must be $16.5 \times 800,000$ or 13,000,000 light-years away.

In order to find the value of k, then, we must divide Virgo's velocity of recession in miles per second by its distance from us in millions of light-years. We find that $k = 710/13$, or about 55.

We can expect then that a galaxy that is 1,000,000 light-years away should be receding from us at a velocity of 55 miles per second; one that is 2,000,000 light-years away should be receding from us at a velocity of 110 miles per second; one that is 10,000,000 light-years away should be receding from us at a velocity of 550 miles per second, and so on.

Is there any limit to this, supposing Hubble's law to hold exactly at all distances? Mathematically, there is not. A galaxy that is 1000 million light-years away should recede at a velocity of 55,000 miles per second; one that is 1,000,000 million light-years away should recede from us at a velocity of 55,000,000 miles per second, and so on.

Physically, however, there *is* a limit. Einstein's "special theory of relativity" (advanced in 1905, ten years before the more all-encompassing general theory) makes it necessary to suppose that the maximum velocity that can be measured relative to one's self is the velocity of light in a vacuum. This is equal to 186,282 miles per second (or 299,776 kilometers per second).

At a certain distance, a galaxy must be receding from us at that velocity and, according to the viewpoint of relativity, that represents an absolute, insurmountable limit. We can deal with nothing moving more rapidly than that, and therefore with nothing more distant from us than that.

There might be all sorts of arguments concerning whether or not there were galaxies farther away, but that would all be irrelevant. It would not matter if there were or not. The point is that once we reach a point so distant from ourselves that a galaxy recedes at the velocity of light in a vacuum, light from that galaxy cannot reach us. The Doppler-Fizeau principle stretches out each wavelength infinitely and therefore reduces its energy to zero. Nothing reaches us from a galaxy that distant. No light, no radiation of any sort, no neutrinos, no gravitational influence. *Nothing.*

Even if it were possible to conceive of something beyond that limit of distance, it would be something that would remain forever undetectable to us — not because of the imperfection of our instruments but because of the nature and design of the Universe. Consequently, we can forget all about any talk of a Universe of infinite dimensions. We must talk, instead, about an *observable* Universe that is finite in diameter and volume.

It remains only to tell what the diameter of the observable Universe is. To do this, let us begin with the mathematical equation representing Hubble's law, $V = kD$, and solve for D (distance from us). This changes the equation to:

$$D = V/k$$

What we want is the distance from us that represents a

velocity of recession equal to the speed of light. We set V equal to the speed of light or 186,282 miles per second, and let k equal 55. In that case D (in millions of light-years) equals 186,282/55 or 3400. This means that the limit of the observable Universe seems to be 3400 million light-years — or 3.4 billion light-years — in all directions from us.

To put it more concisely, the observable Universe is, apparently, a sphere with ourselves at the center and a radius of 3.4 billion light-years and, consequently, a diameter of 6.8 billion light-years. The limit of the observable Universe is, in other words 4250 times as far from us as the Andromeda galaxy is, according to this analysis.

These are enormous dimensions and they certainly sound like a fitting climax to man's long search for limits, a search that began with his consideration of the horizon a mere few miles away.

And yet — there were problems. There was something wrong with the scale of the Universe as it was worked out by the 1940's through the use of Hubble's constant.

The Cepheid Yardstick Revised

The distances worked out for the far-off galaxies were based on the comparison of their apparent brightness with that of near-by ones whose distance was in turn determined by the Cepheid yardstick. And of the nearby ones, the distance determination was most certain and reliable, it seemed, for the Andromeda galaxy. If the distance of the Andromeda galaxy was wrong, then all the distances were wrong; the entire scale of the Universe was wrong.

And by 1950, the uncomfortable feeling was growing that the determination of the distance of the Andromeda galaxy was indeed in error. If Andromeda was at a distance of 800,000 light-years, as the Cepheid yardstick seemed to indicate, certain peculiarities showed up. For one thing, the Andromeda galaxy seemed to be considerably smaller than our own Galaxy, perhaps only a quarter as large. There was no crime in this, taken alone, but all the galaxies whose size could be estimated seemed to be considerably smaller than our Galaxy.

One might argue that some one particular galaxy had to be larger than all the others, and we just happen to be living in that one. And yet why should our Galaxy be so *much* larger?

Whatever process formed the galaxies produced them in a

wide range of sizes. No one could argue with the fact that the Galaxy was far larger than the Magellanic Clouds, and that the Andromeda galaxy was far larger than its satellites, M32 or M33. But there were numerous members representing every portion of the range; no single galaxy was unique in size, either at the large or the small end of the scale — except our own. Our Galaxy stood alone, far larger than the rest.

Furthermore, our Galaxy was the wrong shape to be so large. Where different galaxies could be compared directly, it was always the elliptical galaxies — particularly the spheroidal ones classified as E0 — that were the giants. Why should the largest of all, our own Galaxy, be a spiral?

What was worse still was that our Galaxy was not only larger all together, but that its component parts were larger and brighter than the analogous component parts in other galaxies such as the Andromeda.

For instance, the Andromeda galaxy has a halo of globular clusters about its center, just as our own Galaxy has (see page 58). The number of globular clusters, their appearance, and their distribution are all very similar in both cases. One could, however, begin with the apparent brightness of the individual globular clusters of Andromeda and, considering them to be at a distance of 800,000 light-years, work out what their actual luminosity must be. It turns out that the globular clusters of Andromeda are less than a quarter as bright, on the average, as our own globular clusters are, and only about half as wide in diameter. Even individual stars showed the same effect. Ordinary novae, appearing in Andromeda, usually attained considerably less luminosity than novae in our own Galaxy did, allowing for an 800,000-light-year distance.

To suppose that our own Galaxy was not only a giant among galaxies, but that it was made up of globular clusters that were giants among globular clusters, and of stars that were giants among stars, was asking too much. It looked almost as though we were looking at the Andromeda galaxy (and, therefore, at the other galaxies, too) through a diminishing glass that was reducing everything about it in size. Since everything about the Andromeda galaxy was determined on the basis of its distance, the question had to arise as to whether that distance might not be wrong. Since the distance, as accepted in 1950, depended, in turn, on the Cepheid yardstick, the question had to arise as to whether there might not be something wrong with the Cepheid yardstick.

Baade, in the early 1950's, addressed himself to this question. He reasoned that the stars of the Magellanic Clouds and of the globular clusters of our own Galaxy were of Population II (see page 179), the generally smaller and stabler of the two populations. It had been Population II Cepheids, therefore, that had been used to set up the period-luminosity law in the first place, and it had been those which had been used to determine the scale of our Galaxy and the distance of the Magellanic Clouds.

However, the Cepheids that had been used to determine the distance of the Andromeda galaxy (and therefore, indirectly, of all the far-off galaxies) had been those of the spiral arms of the Andromeda because the giant blue-white members of the Population I stars in those arms had been the most easily seen at Andromeda's vast distance. Could it be that the Population I Cepheids of the spiral arms of Andromeda did not follow the same period-luminosity law followed by the Population II Cepheids that Leavitt and Shapley had worked with?

Certainly, there seemed to be considerable difference between the two types of Cepheids. The Population II Cepheids included a considerable number with particularly short periods running from an hour and a half to a day, whereas such periods were quite rare among the Population I Cepheids, where periods of several days to several weeks were much more common. Secondly, the Population II Cepheids were, on the whole, smaller and dimmer than the Population I Cepheids. This second difference was masked by the fact that the Population I Cepheids in our own Galaxy, located in the dusty spiral arms, were dimmed and reddened by the interstellar dust by an amount that had not been properly allowed for.

The Population II Cepheids are, in fact, even in possession of a special name because of the distinctiveness of their properties compared with other variables. They are the "RR Lyrae stars," named for RR Lyrae, the first (and nearly the brightest) of these variables to be studied. Because RR Lyrae stars are regularly found in globular clusters, they are sometimes called "cluster-type variables."

Baade carefully studied the Population II Cepheids and the Population I Cepheids separately, and in September 1952, announced that the period-luminosity law as worked out by Leavitt and Shapley, applied only to the Population II variety. The distance of the Magellanic Clouds and the dimensions of our Galaxy were therefore correct. The Population I Cepheids, however, followed a somewhat different relationship and for a given periodicity were a magnitude or two brighter than would have been expected from the ordinary relationship used by Shapley.

Let us see what this means. Suppose we observe a distant Cepheid with a period that yields us an absolute magnitude of −1. This means that if it were 32.5 light-years (10 parsecs) from us, it would appear to have a magnitude of −1. To be reduced from −1 to its actual magnitude of something like 20, it would have to be some 24,000 times more distant than 32.5 light-years — or 800,000 light-years away.

But suppose it turned out that, using Baade's new period-luminosity scale for Population I Cepheids, the particular Cepheid under study had an absolute magnitude of — 3 rather than —1. It would then be more than six times as bright as had been thought.

The two populations of Cepheids

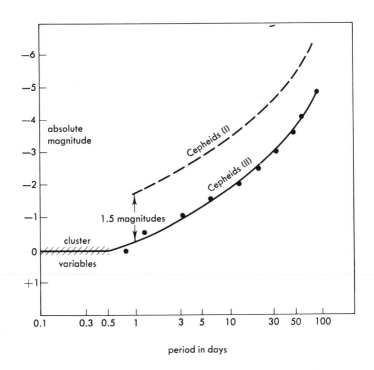

period in days

In order to reduce such a six-times-brighter star to a magnitude of about 20, it would have to be placed correspondingly farther off — 58,000 times more distant than 32.5 light-years, or nearly 2,000,000 light-years away.

By using the revised Cepheid yardstick and adding some additional refinements that now appear necessary, the Andromeda galaxy is today thought to be some 2,700,000 light-years away. All other galaxies beyond the Andromeda must be moved correspondingly.

This removed, at once, all the uncomfortable uniqueness of our Galaxy. If Andromeda is 2,700,000 light-years away (rather than 800,000) and still appears as large and as bright as it does in a telescope, it must be much larger and brighter in actual fact than had been supposed in the days when the shorter distance was accepted.

Nowadays, the Andromeda galaxy is accepted as being somewhat *larger* than our Galaxy. The Andromeda is as much as 200,000 light-years across and contains as many as 300,000,000,-000 stars. Moreover, its globular clusters, which are also farther away than had been thought, are now seen to be larger and brighter than they had been considered — as large and as bright, in fact, as our own globular clusters. The novae in the Andromeda are also as large and as bright as those in our own Galaxy. Furthermore, all other galaxies are now seen to be larger and brighter than had been thought, and many of the spirals rival our Galaxy in size, while some spheroidal galaxies may be ten to thirty times as large.

Our Galaxy remains a giant galaxy, but it is no longer unique, no longer a one-of-a-kind monster. The 135-billion-star Milky Way Galaxy fits well in a Universe that contains galaxies varying in star number from 10- to 5000-billion.

Since this new scale of distance has removed the most serious peculiarities from the galactic scene, astronomers are hopeful that they now have the scale about right. Certainly in the years that have passed since Baade's correction, nothing has happened to shake this faith. In fact, since Baade's death in 1960, astronomers such as the Russian-American Sergei Illarionovich Gaposchkin (1898–   ) have continued analyzing the photographs of Andromeda taken by Baade, using the 200-inch telescope, and have confirmed his work completely.

The new scale of distance has not, of course, affected the red-shift determinations. These determinations are independent of distance. The Virgo cluster of galaxies is receding from us at a rate of 710 miles per second whatever distance we determine for it. From the brightness of its brighter members, as compared with the brightness of the Andromeda galaxy, it is still 16.5 times as far away as the Andromeda.

But now that the accepted distance of Andromeda has been tripled, so must the accepted distance of the Virgo cluster. It must now be considered at a distance of 2,300,000 × 16.5 light-years or something like 38,000,000 light-years away, rather than merely 13,000,000.

To determine Hubble's constant, we divided the velocity of recession of a galaxy or cluster of galaxies by the number of millions of light-years it is distant from us. Instead of dividing 710 by 13, we must now divide it by 38, so that Hubble's constant comes out to be 18.5 rather than 55. If anything, this still probably errs on the side of conservatism. Let us therefore set the value of Hubble's constant at 15.

To determine the distance at which a galaxy must be receding at the speed of light, let us once again use the equation: $D = V/k$, setting V equal to 186,282, and k, this time, at 15. It turns out that D equals 12,500 and we can therefore say that a galaxy at a distance of 12,500 million light-years, or 12.5 billion light-years, can no longer be detected. That is the limit of the observable Universe, or the "Hubble radius."

To put this another way, we can say that the diameter of the sphere of the observable Universe (with ourselves at the center) is 25 billion light-years — a diameter nearly four times that thought correct as late as 1950.

15

The Beginning of the Universe

The Big Bang

The change in the scale of distance of the Universe did more than remove the anomaly of our Galaxy's apparent super-gianthood. It greatly lessened an even more serious discrepancy.

In the second quarter of the twentieth century, astrophysicists and geologists once again disagreed on the age of the Earth, as they had done nearly a century earlier in Helmholtz's day (see page 105).

Again the discrepancy arose over a phenomenon that seemed to cause no trouble in the present and in the future, but raised serious difficulties when extrapolated into the past. In Helmholtz's day, it had been the supposed contraction of the Sun; in Hubble's day, it was the expansion of the Universe.

If one tries to look forward in time, then, and if one accepts the fact that the galaxies will continue indefinitely to recede from each other in the present manner, no insuperable difficulties arise. Every galaxy outside our Local Group will continue to recede at a regularly increasing rate, matching its regularly increasing distance. The galaxies will grow dimmer and dimmer, both because of their increasing distances and because of their increasingly pronounced red shifts and consequent decreasing light energies. Eventually all will approach the limit of the observable

Universe and be lost to us. The observable Universe will then consist only of our Local Group.

This sounds like a lonely future, but we will lose only objects not visible to the naked eye, objects of whose existence and true nature we only became aware of in the last fifty years. The loss, therefore, is not a great one to non-astronomers. Furthermore, it will not happen for a long time—a hundred eons or more—and by that time, events more immediately affecting us will have occurred. Our Sun will have become a white dwarf and our planetary system will be uninhabitable, even if we suppose it to have survived the Sun's red giant stage. All stars larger and brighter than the red dwarfs will certainly be white dwarfs, and all the galaxies will be in extreme old age. New stars may have formed, but a hundred eons from now the supply of dust and gas may have sunk to minimal values and such new stars may be few indeed. Furthermore, the final new stars may be formed out of gas so charged with complex atoms (which will have been spread through space by the hundreds of millions of supernovae that will have exploded in the interval) that they will have abnormally low hydrogen supplies and will be particularly short-lived.

Yet if this future sounds grim, it does not seriously conflict with any accepted scientific beliefs and poses no serious dilemma for astronomers. The Universe cannot be expected to respect human emotions. It can age and die without regard for man's regrets, and its large components can continue to recede from each other in eternal expansion even after the galaxies have flickered down to white dwarf cinders.

But suppose we look backward in time. Suppose we run the expansion of the Universe in reverse, as though it were a movie film. In that case, we must picture the various galaxies approaching each other at known rates, and this time the process cannot be continued eternally. Eventually, the galaxies must meet. If Hubble's law holds, so that every galaxy moves inward at a velocity proportional to its distance from any particular galaxy (such as our own) used as standard, then the Universe generally must be looked on as contracting and all the galaxies must meet simultaneously at a point.

At some particular time in the past, then, all the matter and energy of the Universe must have existed in one large lump. At that time in the past ("zero-time") the Universe could not possibly be as it appears today; the Universe-we-know could only have existed since this zero-time and this zero-time can in fact be considered the beginning of our Universe.

It is possible to calculate this zero-time of the Universe from the distances separating the galaxies now and the rate at which the Universe is now expanding. According to the scale of distances accepted from 1925 to 1952, this zero-time must have been approximately 2,000,000,000 years ago.

Two eons is a long time; it is certainly longer than the twenty million years allotted the Earth by Helmholtz. However, 2 eons is still not long enough for the geologists and their dismay was great. What was the use of saying that the Universe existed as a single glob of matter 2 eons ago and that all the galaxies had formed since then, when the Earth itself was found to be more than twice as old as that on the basis of uranium-lead determinations (see page 109).

The Earth simply could not be twice as old as the Universe. Something was seriously wrong either with the uranium-lead ratios or with Hubble's constant.

It was a standoff until Baade's work showed that the trouble lay with the scale of distance of the Universe and that this had previously given astronomers far too high a value for Hubble's constant. Once again the geologists were right and the astronomers had been wrong. The new scale of distance of the Universe made it appear that zero-time is more like 15 eons in the past.

However, no one pretends that the determination of Hubble's constant is beyond correction even today. In 1979, three Harvard astronomers presented evidence that had led them to believe that the Universe was expanding more rapidly than had been thought and had reached its present extent at a rate that made zero-time only 9 eons in the past. Majority opinion is likely to remain at 15 eons, however, at least for a while.

But what was it that happened at zero-time? The first to consider that point effectively was the Belgian astronomer Georges Edward Lemaître (1894–1966). In 1927, he suggested that at zero-time, the matter and energy of the Universe were actually and literally squashed together into one huge mass, perhaps no more than a very few light-years in diameter. He called it the "cosmic egg" because out of it, the "cosmos" (a synonym for "universe") was formed.

The cosmic egg was unstable and exploded in what we can only imagine to have been the most gigantic and catastrophic explosion of all time, for the fragments of that explosion became the galaxies, which were sent hurtling out in all directions. The effects of that explosion are still with us, for we see it as the recession of the galaxies and clusters of galaxies from each other.

If the different fragments of the cosmic egg were hurtled outward at different velocities (depending on where in the egg the fragment was originally located, and how much it was slowed down by collisions with other fragments), those that ended with high velocity would naturally gain constantly on those that ended with low velocities. This, Lemaître maintained, would give rise to the situation we now experience: one in which the galaxies are receding from each other, with the rate of recession proportional to distance. (It is also possible that even if the cosmic egg had no angular momentum, some of its fragments would pick up clockwise angular momentum as a result of the explosion and other counterclockwise angular momentum, the sum adding up to zero.)

The Lemaître model of the Universe is a physical analogue of Sitter's theoretical model. Sitter's Universe expanded simply because that fit a set of equations worked out by Einstein. The Lemaître model, on the other hand, expanded in consequence of a physical event; an explosion that differed in size, but not in nature, from that of a firecracker on Earth. The Lemaître model is easily grasped; it is concrete, dramatic, and seems familiar. Eddington adopted and popularized it and since his time, the Russian-American astrophysicist George Gamow (1904–1968) has upheld it enthusiastically. With reference to that vast initial explosion of the cosmic egg, Gamow termed the Lemaître model of the Universe, the "big bang theory," although it might less dramatically be called the "exploding Universe theory."

Naturally, one is curious as to the nature of the cosmic egg. What was it made of? What were its properties?

Perhaps we can get an idea if we try to observe the Universe (in imagination) running forward and backward in time. Right now the Universe seems to be roughly 90 percent hydrogen, 9 percent helium, and 1 percent more complex atoms. As the Universe runs forward in time, hydrogen continually fuses to helium, and helium to still more complicated atoms within stellar cores (see page 152). If we run the Universe backward, the quantity of helium and more complex atoms decreases and the quantity of hydrogen increases. As we approach zero-time, then, we should expect the Universe to consist entirely, or almost entirely of hydrogen.

But the matter and energy of the Universe are compacting themselves as we view matters backward in time. At zero-time, the hydrogen that exists must ultimately be compressed; all the particles composing it are pressed together as hard as they can be.

The hydrogen atom is made up of two particles only, a central proton carrying a positive electric charge and an outer electron carrying a negative electric charge. As long as these exist separately, there is a limit to how compressed a mass of hydrogen can be. If some critical pressure is surpassed, however, the electrons and protons may be considered as squashing together to form a mass of electrically uncharged particles called neutrons.

Such a mass of ultimately compressed neutrons is sometimes

The Big Bang Theory

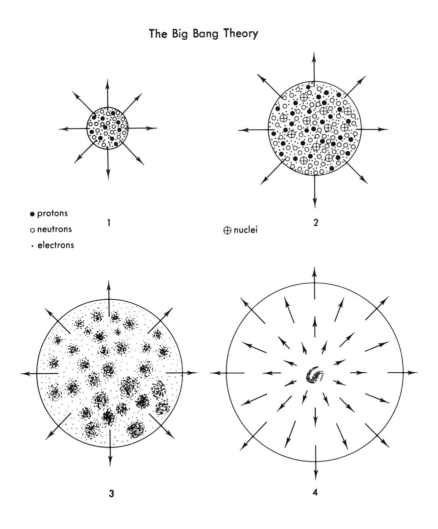

• protons
o neutrons
· electrons

⊕ nuclei

1

2

3

4

called "neutronium" (although Gamow uses the term "ylem," a Latin word for the substance out of which all matter is formed). It would have a density of about 1,000,000,000,000,000 grams per cubic centimeter, and would be far denser than the densest known white dwarf.

The Formation of the Elements

Without committing himself on the nature of the cosmic egg, Lemaître had viewed it as a kind of radioactive superatom, one which would break down as ordinary radioactive atoms do but on an incredibly larger scale. Not only did portions of the cosmic egg form the present-day galaxies but, on a more intimate scale, the cosmic egg broke up to form the atoms we know today. From the Lemaître viewpoint, however, the atoms would form from the top down, so to speak. Very massive atoms would be formed and these would break down further, producing less massive atoms and so on, progressively, until permanently stable atoms were formed. This, however, would mean a Universe composed chiefly of such atoms as lead and bismuth, which are the most massive stable atoms in existence. It would not account for the preponderance of hydrogen in the Universe.

An alternate view as to the formation of the elements was offered by Gamow, who presented it in 1948 in association with Bethe and with the American physicist Ralph Asher Alpher (1921–   ).[1]

According to Gamow's suggestion, the neutronium cosmic egg, at the instant of the big bang, disintegrated with ferocious violence into separate neutrons, which rapidly broke down into protons and electrons. (Individual neutrons do this today for that matter, the half-life of the breakdown being about thirteen minutes.) The protons formed can be considered the nuclei of hydrogen-1 atoms.

As the protons formed, they would sometimes collide with neutrons that still persisted and gradually build up additional stable atomic nuclei of greater complexity. The advantage of this theory is that it makes use of the phenomenon of neutron-addition, some-

[1]   Gamow is supposed to have selected these associates with puckish humor, for the names Alpher, Bethe, and Gamow resemble the first three letters of the Greek alphabet, alpha, beta, and gamma. Using the Greek symbols for these letters, the theory is sometimes referred to as the α, β, γ theory.

thing which atoms are prone to do and which can be observed in the laboratory.

If a proton combined with a neutron, for instance, it would form a nucleus of hydrogen-2, or deuterium (one proton/one neutron). Hydrogen-2, combining with another neutron, would form hydrogen-3 or "tritium" (one proton/two neutrons). Tritium is, however, unstable. One of the neutrons in its nucleus emits an electron and becomes a proton, and the nucleus becomes helium-3 (two protons/one neutron). The helium-3 nucleus adds a neutron to become the common helium-4 (two protons/two neutrons). The process continues and gradually, one neutron at a time, the whole list of elements is built up.

At the incredibly high temperatures following the explosion of the cosmic egg, Gamow envisages the necessary nuclear reactions to take place very rapidly, even perhaps within the first half-hour. Gradually, thereafter, as the temperatures dropped, the various nuclei would attract electrons to themselves and form atoms; the atoms would conglomerate into huge volumes of gas speeding outward from the site of the exploded cosmic egg and gradually condensing into galaxies and stars as they sped along.

Naturally, only a small portion of the hydrogen-1 nuclei first formed would undergo collision with neutrons to form hydrogen-2; only a small portion of the hydrogen-2 nuclei would undergo a further neutron collision to form helium-3, and so on. Each successively more complex atom would be less common than the one before, and this would account for the fact that in the Universe today there is a more or less steady drop in the abundance of atoms with a rise in complexity.

The drop is not an absolutely uniform one. Helium-4 is much more common than either hydrogen-2 or helium-3, and iron-56 is much more common than most of the atoms less complex than itself. On the other hand, simple atoms such as those of lithium-6, lithium-7, beryllium-9, boron-10, and boron-11, are less common on a cosmic scale than they ought to be, considering their simplicity. The Gamow theory can make a stab at explaining this. Helium-4 and iron-56, for instance, are both examples of particularly stable nuclei. They would react to form more complex atoms only with difficulty and would therefore pile up. The atoms of lithium, beryllium, and boron, on the other hand, react particularly easily and would "burn up."

Gamow's theory would account for the relative occurrence of the different atoms in the interstellar material. Once stars form, other changes take place in their cores.

Gamow's theory has, however, one serious flaw which no one, so far, has managed to argue away. The atoms must be formed one neutron at a time, and there is a gap that cannot be leap-frogged once the helium-4 nucleus is reached. The helium-4 nucleus is so stable that it has virtually no tendency to accept either a neutron or a proton. If a neutron does manage to attach itself to the helium-4 nucleus, it forms a helium-5 nucleus (two protons/three neutrons), which breaks down in about 0.0000000000000000000001 seconds (a thousandth of a billionth of a billionth of a second) to form helium-4 and a single neutron again. On the other hand, if a proton manages to attach itself to helium-4, lithium-5 (three protons/two neutrons) is formed and that breaks down to helium-4 again even more quickly.

Suppose, on the other hand, a helium-4 nucleus is struck by another helium-4 nucleus and the two fuse. This is a much less likely occurrence than the fusion of helium-4 with the very common individual protons and neutrons, and even so it is useless as a way out. Beryllium-8 is formed and that breaks down to two alpha particles with super-rapidity too.

In other words, once you have formed helium-4 by neutron addition, you are stuck. There is a gap at 5 and another at 8 that seem insurmountable.

It is possible, of course, that two particles may strike the helium-4 nucleus simultaneously. If a proton and neutron both strike and attach themselves, then lithium-6 (three protons/three neutrons) is formed. That is stable and the process can then continue.

Unfortunately, under the conditions postulated by Gamow in the first half-hour following the big bang, the individual nuclei are so widely dispersed that the chance of two particles striking the alpha particle simultaneously is virtually zero. The Gamow model, then, seems to account only for hydrogen and helium atoms but nothing beyond that.

Opposed to this theory of the formation of the elements is one that I have implicitly accepted in this book and used in my discussion of second generation stars (see page 170). This theory is suggested by Fred Hoyle, who considers that hydrogen-1 only is the original material and that everything else is formed within stars and is added to interstellar material by way of supernovae.

Hoyle makes use of the same mechanisms proposed by Gamow, but now there is a difference. In the stellar core, the density of matter is much higher than in open space. The chance of a helium-4 nucleus being struck by two particles in an essentially simultaneous manner is therefore considerably better than in the Gamow theory. In fact, since the stellar core is richer in helium-4 than in anything else, there is a reasonably likely chance that a helium-4 nucleus will be struck by two other such nuclei in sufficiently rapid succession to form a carbon-12 nucleus. This would bypass the stable atoms between helium-4 and carbon-12—the lithium, beryllium, and boron atoms previously mentioned. Those light atoms would be formed only by less common secondary processes and that would account for their relative rareness in the Universe today.

The formation of the elements in stellar cores not only avoids the gap at the 5-particle and 8-particle levels but is also favored by an interesting piece of evidence. The spectrum of certain unusual stars of spectral class S shows evidence for the presence of an element called "technetium." Technetium is a radioactive element that possesses no stable variety of atom. The most nearly stable variety is technetium-99 which has a half-life of about 220,000 years. This is long on the human scale, but after five million years (no time at all in the life of an ordinary star) only a billionth of any original technetium-99 would remain. It follows that if technetium can be detected spectroscopically now, it cannot have existed at the time the star was formed but must have been made fresh, so to speak, within the star's interior.

All in all, then, the weight of plausibility and of what evidence there is, seems, at the moment, to favor the Hoyle model of element formation over the Gamow model.

Before the Big Bang

If we postulate the existence of a cosmic egg marking the original form of the Universe, with its explosion as zero-time, we are bound to ask: But where did the cosmic egg come from?

We can avoid having to answer that by seeking refuge in eternity. The law of conservation of energy implies that the substance of the Universe is essentially eternal, so we can say that the matter making up the cosmic egg was always there.

Yet, even granting that the matter of the cosmic egg was always there, was it always in the form of the cosmic egg? If the cosmic egg, as such, had always existed, it would have to be stable. If it were stable, why did it suddenly cease being stable and explode at what we call zero-time, after uncounted eons during which it had not exploded but had merely existed?

We would face the same problem on a merely stellar scale if we asked why a star should explode into a supernova after having existed for eons under conditions of reasonable stability. In the case of a star, however, we have learned enough to account for this in terms of progressive nuclear reactions proceeding in the stellar core.

Unfortunately, we cannot study a cosmic egg; we have no knowledge of what can go on within it; we do not know what forces will suffice to keep it stable or how they will bring about progressive changes that will eventually and suddenly make it unstable.

If we ask ourselves, however, in what form the substance of the Universe might exist in order to remain stable over countless eons, without our having to strain our imaginations to account for the stability, we might find it easiest to think of the Universe as an exceedingly thin gas. The Universe would then be the kind of "empty space" that now exists between the galaxies and that is certainly stable.

Such an exceedingly thin gas would still be subject, however, to its own vastly diffuse gravitational field. Slowly, over the eons, the gas would collect and the Universe would draw closer together. As the substance of the Universe grows more compact, the gravitational field becomes more intense until, after many eons, the Universe is contracting at a great rate.

This contraction must heat the Universe, however, à la Helmholtz, and produce a higher and higher temperature in matter compressed into a smaller and smaller volume. The temperature rise increasingly counters the gravitational contraction and begins to slow it down.

The inertia of matter keeps it contracting, however, past the point where the temperature effect would just balance gravitation. Finally, the Universe contracts to a minimum volume, represented by the cosmic egg or a close approach to it. At some point of the outward push of temperature and of radiation finally gains control, and the substance of the Universe is pushed outward, faster and faster, in a manner that rapidly builds up to the big bang.

In this view, the Universe starts in a state of virtual emptiness, goes through a phase of contraction to maximum density, and then through a phase of expansion to emptiness again. We do not need to puzzle ourselves over a cosmic egg that existed "to begin with" and then, after an indefinite period of stability, suddenly exploded.

Instead, the cosmic egg becomes a momentary object placed midway in eternity.

This model is termed the "hyperbolic Universe." It can be pictured graphically by considering its "radius of curvature." A ray of light traveling endlessly through the kind of Universe pictured by Einstein as having a Riemannian geometry (see page 190) would describe a vast circle the radius of which would be the radius of curvature of the Universe. In a contracting Universe, this radius would be decreasing; in an expanding Universe, it would be increasing. In a hyperbolic Universe, it would first decrease to a minimum and then increase again.

The hyperbolic Universe lasts through eternity, but it is not truly eternal in the sense that it persists always in an essentially unchanged condition, or in a condition that hovers about some unchanged average. Instead, it undergoes a permanent and irreversible change. It begins as an empty Universe filled with a thin gas, presumably hydrogen. It ends as an empty Universe filled with innumerable white dwarfs. There is a definite beginning and a definite end, and we inhabit the brief interval of time during which the Universe deviates for an instant from its eternal emptiness.

However, the hyperbolic Universe is not the only possible model one can deduce from a consideration of the cosmic egg. If the Universe is pictured as being blown into outward-hurtling pieces by the force of a gigantic explosion, there still remains the

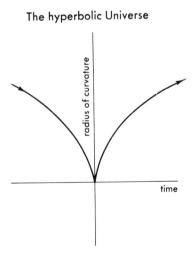

The hyperbolic Universe

force of universal gravitation acting to pull the pieces back together again, and conceivably, it might, succeed.

To understand what this means, let us consider a similar situation on Earth. An ordinary explosion may hurl objects forcefully
into the air, but the velocity at which they recede from the Earth
is continually decreasing because of the pull of gravity. The objects eventually come to a momentary halt and then begin to fall
back to Earth. The more forcefully the object is hurled upward in
the first place, the higher it will travel before halting and returning.

The Earth's gravitational field weakens with distance, however, and if the object is hurled upward forcefully enough, it
reaches regions where Earth's steadily weaking gravitational field
will never quite suffice to bring the object's steadily diminishing
velocity all the way to zero. The object will have been shot upward
at more than the "escape velocity" which, from the Earth's surface,
is just about seven miles per second.

Without knowing the actual size of the cosmic egg, or its mass,
or the force of the explosion that rent it, it is difficult to determine
whether the fragments flying outward from it managed to attain
escape velocity or not. Are the galaxies receding from each other
forever, or will their velocity of recession slowly decline with time,
reach a momentary zero mark, and then will the galaxies finally
begin to fall together again—very slowly at first, but then more
and more rapidly?

Suppose the galaxies do someday begin to fall together again.
In such a contracting Universe, the radiation emitted by the galaxies undergoes a violet shift and the extent of that violet shift increases as the velocity of approach grows greater with accelerating
contraction. The energy pouring into the center of the Universe is
compressed, so to speak, and heightened. Under the lash of that
energy outpouring, the nuclear reactions that take place in an expanding Universe are reversed.

Where fusion from hydrogen to iron would yield energy in
an expanding Universe of generally dimming radiation, breakdown
from iron to hydrogen would absorb energy in a contracting Universe of generally brightening radiation.

In short, by the time the Universe was condensed to something approaching the cosmic egg, it would be all hydrogen again.
Following the formation of the cosmic egg there would be another
big bang and the whole procedure would start over again. The result would be an endlessly "pulsating Universe" or "oscillating
Universe."

A pulsating Universe might be looked on as a gigantic Cepheid variable. Such a Universe would be eternal in a real way, for although catastrophic changes take place, these are periodic. There is no clear beginning or ending; no steady, irreversible change from one grand universal structure to another quite different one. As the Universe is right now, so it will be again an indefinite number of eons in the future after it collapses and explodes again, and so it was an indefinite number of eons in the past prior to its last collapse and explosion.

Continuous Creation

Yet prior to 1952, at least, the big bang theory seemed to have an element of impossibility to it. It placed zero-time 2 eons in the past, when the Earth was nearly 5 eons old. Somehow the big bang had to be an illusion, and a model of the Universe had to be constructed which did not involve the cosmic egg at all.

The new model arose out of the feeling that the general scheme of appearance of the Universe would be the same from any vantage point. No matter where an observer might be placed in the Universe, no matter on which galaxy, he would find all the

The pulsating Universe

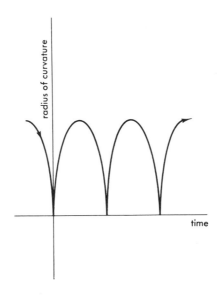

galaxies distributed symmetrically about him in all directions; he would find the general density of matter the same; he would find all the other galaxies receding at a rate proportional to their distance; he would find himself the center of an observable Universe.

This notion of a Universe generally uniform through all space was advanced by the English cosmologist Edward Arthur Milne (1896–1950). He termed it the "cosmological principle."

The cosmological principle is only an assumption but, barring strong evidence of its falsity, astronomers are attracted to it, since by using it, the Universe can be made simple enough to be reflected in the kind of models astronomers can construct. Einstein, for instance, assumed the cosmological principle when he treated the Universe as though the matter in it were smeared out evenly, for then the Universe certainly seems the same no matter where you are located in space.

The cosmological principle seems to require an infinite Universe, for otherwise you could imagine yourself transported to the very edge, where you would find all the galaxies on one side and nothing at all on the other. How would this jibe with the Riemannian Universe assumed by Einstein, a Universe finite in volume?

Actually, it is possible to have a Universe that is finite in volume and yet contains an infinite number of galaxies.

By Einstein's theory of relativity, it is necessary to suppose that an object moving with respect to ourselves is found to be, by any measurement we can make, shorter in the direction of its motion than it would be if it were motionless with respect to ourselves. The greater the velocity, the more pronounced this "foreshortening." If the object is moving at the velocity of light, its length in the direction of motion is reduced to zero.

The distant galaxies, receding from us, must be foreshortened to our view; and the more distant, the more foreshortened, in view of the increasing velocity of recession with distance. Near the edge of the observable Universe, the galaxies are paper-thin and less, and an infinite number can be squeezed into the very rim. We then have an infinite Universe packed into an infinite volume. (Because of the red shift, the infinity of galaxies at the rim would deliver only a finite amount of radiation, particles, or gravitational force to the interior.)

An observer on one of the galaxies of the rim would not, of course, find himself and his galaxy paper-thin. His galaxy would seem normal to him as would the other galaxies in his vicinity. At distances far removed from himself, he, too, would observe an infinitely crowded rim, and to him our Galaxy, if he could observe it, would seem paper-thin. (It is a matter of viewpoint, just as the Australians seem upside down to us and we seem upside down to them, if we imagine ourselves looking through a transparent Earth.)

Such an infinite Universe does not fit well with the notion of the cosmic egg, for certainly it is easier to view the cosmic egg as possessing a finite size and as exploding into a finite number of galaxies. Gamow, however, is ready to consider a cosmic egg of infinite size, and in that case the cosmological principle would not be inconsistent with either the hyperbolic or pulsating Universe.

To three astronomers in England, Austrian-born Hermann Bondi (1919–   ), Thomas Gold (1920–   ), and Fred Hoyle, it seemed that the cosmological principle was incomplete. It allowed the Universe to be unchanged with the observer's position in space, but what about his position in time?

If the Universe undergoes changes that are irreversible as in the case of the hyperbolic Universe, or reversible only after many eons as in the pulsating Universe, then an observer would find the Universe varying its nature radically with time. An observer 10 eons ago, for instance, might observe a small filled Universe with young closely spaced galaxies consisting almost exclusively of young stars made up of hydrogen and virtually nothing else. An observer 50 eons in the future might observe a vast empty Universe with galaxies separated by enormous distances and made up largely of white dwarfs. An observer a 100 eons in the future might observe a contracting Universe.

Bondi, Gold, and Hoyle believed it logical to suppose that this could not be so. The Universe would have to be essentially the same for observers at all times as well as in all places, and this they called the "perfect cosmological principle."

Yet the Universe was changing in two important ways; ways that were accepted on the basis of the strongest evidence and with which it was impossible to quarrel. First, the distance between the galaxies is growing steadily larger, and secondly, hydrogen is steadily fusing into helium and more complicated atoms. If the perfect cosmological principle is to be valid, there must be processes that neutralize these changes.

The solution advanced by the three astronomers in 1948 was to suppose that hydrogen was continually being created out of nothing, and this suggestion is referred to as the "continuous creation theory" or the "steady state theory."

Naturally, one's first reaction to any such suggestion is to object that it violates the law of conservation of energy. Yet that law is merely an assumption based on the fact that mankind has never observed energy to be created out of nothing. But the requirements of the continuous creation theory are small indeed; matter need be created only at the rate of one atom of hydrogen per year in a billion liters of space, and such a rate of creation would be far too small to be detectable by any instruments we possess. Such continuous creation would not violate the law of conservation of energy which does not really say "Energy cannot be created out of nothing," but merely "Energy has never been observed to be created out of nothing."

(An alternate view might be that the matter formed appears at the expense of the energy of expansion of the Universe, which would thus expand a bit more slowly than it would if continuous creation were not taking place.)

If continuous creation is allowed, let us next see the consequences. The galaxies must be viewed as separating not as a result of some explosion but as a consequence of some more subtle effect. In 1959, Hermann Bondi and Raymond Arthur Lyttleton speculated, for instance, that the positive charge on the proton might be very slightly larger than the negative charge on an electron. Suppose the positive charge of the proton were only larger by a billionth of a billionth of the size of the electron's negative charge. This would be far too little to detect with man's most refined instruments. It would, however, suffice to build up a generally positive charge on all the galaxies and would cause them to undergo a steady mutual recession. This explanation of the expanding Universe is considered quite unlikely by astronomers generally, but it is an example of the sort of physical cause, other than explosion, that is sought for by those who wish to avoid the big bang.

As the galaxies recede from each other, whatever the cause, the spaces between gradually accumulate matter through continuous creation.

The accumulation is slow, to be sure, but so is the rate at which galaxies recede from each other, compared to the vast spaces between. It takes several eons for the distance between two neighboring galaxies to double, and by that time enough matter has been formed between them to condense into a new galaxy. The density of galactic distribution never grows less, therefore, as the old galaxies spread apart, slowly collecting in the paper-thin rim or, as Hoyle seems to assume, moving beyond, somehow, the rim of the observable Universe. New galaxies would form between them, and the two effects would just balance each other.

Then, too, the matter that is formed in continuous creation would naturally be simple. A fragment of matter would be formed as a hydrogen atom, perhaps, or as a neutron that would break down in a matter of minutes to a proton and electron, which would then come into association to form a hydrogen atom. In either case, the new galaxies formed out of newly created matter would be young galaxies built up out of fresh hydrogen. This means that any observer at any time in the future would see as many galaxies about him as he does now and as many young galaxies among them. The Universe would never become either empty or old although individual galaxies can age to any extent.

If we look backward in time, the galaxies can be pictured as contracting, but they need never come together. Continuous creation in a time-run-backward becomes continuous destruction. Complex atoms break down to hydrogen in such a backward Uni-

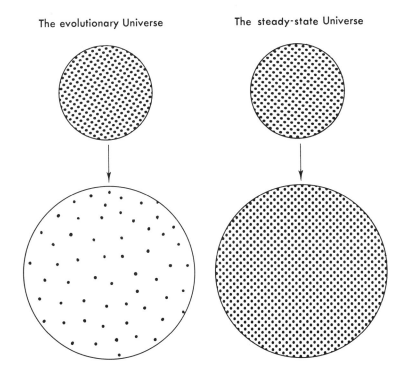

The evolutionary Universe   The steady-state Universe

verse and the hydrogen disappears. Galaxies melt away as they approach each other and never form a cosmic egg. Other galaxies take their place from the infinite supply at the rim or from the infinite supply beyond the rim, depending on one's viewpoint. In the long run, then, galaxies neither get closer together nor do they get younger, however far back in time we go.

Under such conditions, the overall state of the Universe does not change with time in either direction but is steady. Such a model represents a "steady-state Universe" and adheres to the perfect cosmological principle.

There is something very attractive about the thought of an eternal and immortal Universe in which man (or his evolved descendants) might conceivably live forever, and it has appealed very strongly to the general public. This appeal has been strengthened by the fact that the most assiduous proponent of continuous creation has been Fred Hoyle, a charming and persuasive writer, whose popular works on astronomy have been well received by the general public. (On the other hand, George Gamow, who has been the most prominent of those on the side of the big bang, is also a highly successful writer of popularized science. Rarely in the history of science has there been such a clash of titans in full view of the lay public.)

To be sure, in 1952, when Baade proposed a new scale of cosmic distances and shoved zero-time for the big bang back to 6 eons or more in the past, the greatest argument in favor of the steady-state Universe (the argument that the big bang was impossible) vanished. By that time, however, the steady-state Universe had proved too attractive to abandon easily.

It is quite difficult to decide among the models of the Universe presented in this chapter.

To make the decision, we must remember that the steady-state Universe adheres to the perfect cosmological principle, while the others do not. This means that if we could change our position in time, we could solve the problem. If the general appearance of the Universe does not change with time, if the galaxies are not closer and younger in the past or more separated and older in the future, then the steady-state Universe would be strongly supported. Otherwise, either the pulsating Universe or the hyperbolic Universe would be supported, and from the extent of the change with time, we might be able to choose between these two.

To be sure, if our descendants remain in existence for several eons and maintain a continuity of culture, they will be in the far future and will be able to decide, but astronomers would like, if possible, to find the answer *now,* and surely one cannot blame them.

What is needed, then, is time travel, and one form of time travel is possible.

When we say that the Andromeda galaxy is 2,300,000 light-years from us, we mean that it takes light 2,300,000 years to cross the distance that separates it from our eyes. When we look at Andromeda, either with our eyes or by some instrument, we are seeing light that left Andromeda 2,300,000 years ago, and we see it not as it is now, but as it was that length of time ago. In studying the Andromeda galaxy we are, in effect, time travelers who have penetrated 2,300,000 years into the past.

The farther we penetrate out into space, the longer it takes light to reach us from the objects we can see, and the farther back in time we find ourselves. Our best optical telescopes (as far as was known in the 1950's) were reaching objects one or two billion light-years away, and in looking at them we are seeing that portion of the Universe as it was 1 or 2 eons ago.

If the steady-state view of the Universe is correct, this difference in time should not matter. The Universe 1 or 2 eons ago would have the same general properties it has today. The galaxies we see at the far end of our telescopic capabilities should be spaced no more closely or no less closely than they are today, they should be receding from each other at the same rate as today, and they should, in general, have no properties that would distinguish them as a whole from those galaxies we see in our immediate neighborhood.

If the pulsating or hyperbolic views of the Universe are correct, then the difference in time should make for considerable change, and there should be *some* important properties, at least, in which the far-off edge of the Universe should differ from the regions near us.

For instance, the very distant galaxies should be younger than the galaxies in our neighborhood, richer in hydrogen, spaced more closely together, and separating at a greater rate (since the explosive force had not yet been weakend then by the slow, steady pull of gravity). Furthermore, since those regions represent the youth of the Universe, they might include objects not to be found at all in our own neighborhood, objects characteristic *only* of a young Universe. Furthermore, by studying the extent to which these differences make themselves manifest we might determine whether the pulsating Universe or the hyperbolic one better fits the facts.

This seems straightforward enough, but it is actually maddeningly frustrating. The more distant the objects we study, the more likely we are to decide among the suggested models of the Universe; but the more distant the objects we study, the more difficult it is to detect anything at all.

The best that could be done by the middle-1950's was to study the red shifts of the most distant galaxies that could be detected. In a steady-state Universe, Hubble's constant ought to be the same for all times and therefore for all distances. In a pulsating or hyperbolic Universe, Hubble's constant ought to decrease with forward-progressing time and should have been quite high in the youth of the Universe. In that case, the very distant galaxies (representing that youth) ought to be receding more rapidly than

Red shift at great distance

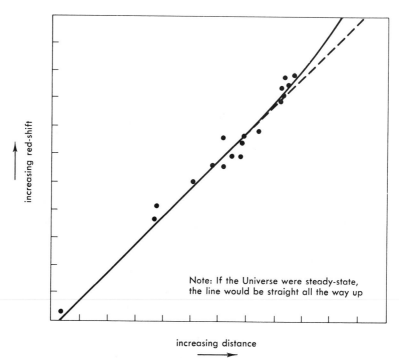

increasing red-shift

Note: If the Universe were steady-state, the line would be straight all the way up

increasing distance

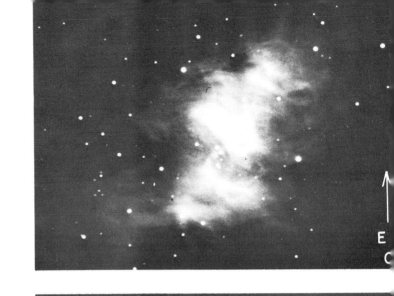

Polarized Light in the Crab
Nebula. *(Photograph from
the Mount Wilson and
Palomar Observatories.)*

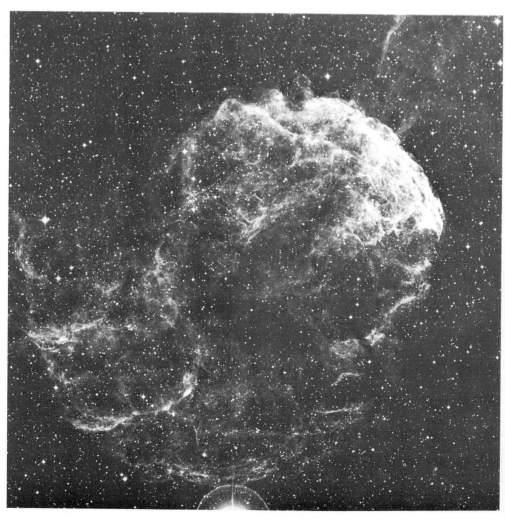

Gaseous Nebula IC 443. (*Photograph from the Mount Wilson
and Palomar Observatories.*)

NGC 5128. *(Photograph from the Mount Wilson and Palomar Observatories.)*

Cygnus A. *(Photograph from the Mount Wilson and Palomar Observatories.)*

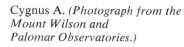

Galaxy M87. *(Lick Observatory Photograph.)*

we would expect and ought to have red shifts greater than we would expect. By 1956, it did indeed seem that such too-great red shifts were observed and this was an argument against the steady-state Universe. However, the red shifts were too large by such small amounts and the difficulty of observing them was so great that the evidence was anything but conclusive. Something better was needed, but what?

To answer that question, we must turn to those branches of modern astronomy that do not depend on visible light. Visible light fails us; other manifestations of the Universe may not.

CHAPTER

16

Particle Bombardment

Massless Particles

All the information we obtain concerning the Universe outside the Earth itself is derived from particles which, emitted by bodies in space, cross the gap and reach Earth. On Earth, they interact with particles already here in such a way that the results of the interaction can be perceived by our senses.

This is a roundabout way of saying that (to give the most common and best-known example) the distant stars and galaxies emit light which we can see, separate by spectroscope, or impress on photographic film.

To be sure, light was viewed as a wave-form, rather than as particles, throughout the nineteenth century. In 1900, however, the German physicist Max Karl Ernst Ludwig Planck, (1858-1947) advanced the "quantum theory" in which light and, indeed, all forms of energy, could be viewed as consisting of discrete little packets which Planck called "quanta" (or "quantum" in the singular). This theory was reinforced by Einstein in 1905, and it became more and more clear that energy quanta could, in some ways, behave like particles.

Indeed, by the 1920's, a kind of duality was recognized. All particles had the capacity to behave like wave-forms in some ways; all wave-forms had the capacity to behave like particles.

231

The two aspects were not ordinarily well-balanced. The particle-like properties of protons, for instance, are much more pronounced and easy to detect than are the wavelike properties, but the latter are there and can be detected if one goes about it carefully enough. On the other hand, the wavelike properties of ordinary light are much more pronounced than are its particle-like properties, but the latter are present also. The shorter the wavelength of light, the greater the energy content of the individual quanta, and the more pronounced and easy to detect are the particle-like properties.

In 1923, the American physicist Arthur Holly Compton (1892-1962), using forms of light of particularly short wavelength, demonstrated their particle-like properties unmistakably. He coined the word "photon" for such particles of light (the "-on" suffix being commonly used for subatomic particles and the "phot-" prefix coming from the Greek word for "light").

The light we see the Universe by, then, can be viewed as a shower of photons descending on us from all directions: from the Sun, stars, and galaxies directly and from the Moon and planets by reflection. The interaction of these photons with the retina of the eye comprised the whole of astronomy prior to the middle of the nineteenth century.

Since then we have expanded matters not only by allowing photons to affect photographic plates rather than the retina directly but by recognizing the existence of and dealing with many forms of particles other than photons. Some of these other forms are subtle indeed, and I will begin with the most subtle of all—particles that resemble photons in some ways.

The photon has a "rest-mass" of zero. That is, if it could be made to stand still, it would turn out to exhibit none of the properties associated with the possession of mass. It would have no inertia, and it would neither produce a gravitational field nor respond to one. It is therefore considered a "massless particle."

Such masslessness is purely theoretical, however, for a photon cannot be made to stand still. The moment it is formed, it moves away from its site of formation at 300,000 kilometers per second.[1]

---

[1]   That is, if it is in a vacuum. If it is traveling through transparent media other than a vacuum, such as air, water, or glass, its velocity is less, sometimes considerably less. Under no conditions, however, can a photon be brought to actual rest without being absorbed. Moreover, the moment a photon passes from some transparent medium back into vacuum, its velocity instantaneously becomes 300,000 kilometers per second again.

While it is moving in this fashion, the photon does exhibit some of the properties associated with mass, that of responding slightly to a gravitational field, for instance.

At least two other massless particles, the "graviton" and the "neutrino," have been postulated by physicists. Both, like the photon, have a rest-mass of zero but are never at rest; they travel only at the speed of light as long as they exist. This would be true, apparently, for all massless particles.[2]

The photon, graviton, and neutrino are all electrically uncharged, and if all three are massless as well, it seems fair to wonder in what manner they may be distinguished. One distinguishing mark stems from the fact that most subatomic particles can be pictured as though they were rotating about an axis, either clockwise or counterclockwise. The angular momentum associated with this rotation can be expressed, therefore, by either a positive or negative number. Physicists use units in such a way as to assign the photon a "spin" of $+1$ or -1. On this basis, in order to account for the manner in which subatomic particles behave, physicists find they must assign a spin of $+\frac{1}{2}$ or $-\frac{1}{2}$ to the neutrino and one of $+2$ or -2 to the graviton. This alone suffices to make the three types of particles absolutely distinct.

The graviton remains a theoretically postulated particle only, for it has never been detected directly. Indeed, such are the properties that the logic of the situation has forced physicists to assign to it, that it may never be detected. Nevertheless physicists suppose that it is by virtue of the emission and absorpion of gravitons that a gravitational field is brought into existence.

If we cannot detect the graviton directly, then we can at least detect it indirectly through the effects of the gravitational fields its produces. Thus, the interchange of gravitons between the Moon, the Sun, and the Earth produces the tides and keeps the Moon and Earth in their interlocked orbit about the Sun. The interchange of gravitons between the Sun and the Galactic center keeps the Solar system in its mighty revolution about that invisible core.

On the whole, the gravitational effect on ourselves of individual stars outside the Solar system, and of individual galaxies beyond our own, is undetectable and will probably continue to be so through the foreseeable future.

---

[2] On the other hand, particles possessing mass, in however small an amount, can never travel at the speed of light. They may approach that speed very closely, but they can never actually attain it fully.

Nevertheless, there are gravitational interactions between neighboring stars and between neighboring galaxies which yield information to us. The two stars of a binary system move about each other, for instance, in accordance with the gravitational law worked out by Newton, and by the application of this law, the relative masses of the two bodies can be determined. Even when one of the members of the system cannot be seen, the motion of the other in response to the gravitons emitted and absorbed by the unseen body, will yield information as to the mass of the latter. That is how Sirius B was first detected (see page 39).

A double star with a component too small to be seen, and detectable only by careful measurements of the tiny wobbles of the component that can be seen ("astrometry"), is called an "astrometric binary." The 20th Century saw far more elusive astrometric binaries than the Sirius system pinned down.

In 1943, the motions of 61 Cygni (see page 40) were studied by a team under the guidance of the Dutch-American astronomer Peter Van de Kamp (1901–   ). The star, 61 Cygni, is actually a binary, so that there is both a 61 Cygni A and a 61 Cygni B, revolving about a common center of gravity. The motion of one of these, however, wavered slightly, but sufficiently to indicate the presence of a body about 1/120 the mass of the Sun. Such a mass, only eight times that of the planet Jupiter, cannot support nuclear reactions to an extent sufficient to qualify it even as a dwarf star. For that reason, the new body, 61 Cygni C, was considered a planet, albeit a giant one — the first planetary body to be discovered outside the Solar system. Others of the sort have been discovered since. In 1963, a close look at Barnard's star (see page 33) showed a slight wavering in its proper motion that indicated the presence of planetary masses. In 1969, Van de Kamp suggested the wavering might be best explained by the presence of two planets, one 1.1 times the mass of Jupiter, the other 0.8 times. Their orbits would be circular. The larger would be the one nearer Barnard's star, about the distance of the asteroid belt from our own Sun, while the smaller would be at a distance equal to that of Jupiter from our Sun. The orbital periods are 12 years and 26 years respectively.

The remaining massless particle, the neutrino, is intermediate between the graviton and photon in ease of detection. Unlike the graviton, the neutrino *has* been detected, but it is by no means as easily detected as the photon.

The existence of the neutrino was first postulated in 1931 by the Austrian physicist Wolfgang Pauli (1900–1958), out of a necessity to explain certain interactions of subatomic particles that otherwise could not be explained. For a quarter of a century, it remained a theoretical figment of the scientific imaginations, and then, in 1956, two American physicists, Clyde Lorrain Cowan, Jr. (1919–   ) and Frederick Reines (1918–   ), designed a careful experiment that clearly demonstrated the very occasional interaction of a neutrino with a proton. The actual existence of the neutrino was accepted at once by the scientific community.

The neutrinos detected in 1956 were produced by uranium fission within the core of a man-made nuclear reactor. Neutrinos (of a somewhat different type) are also produced, and in vastly greater numbers, of course, in the cores of stars (see page 169). Neutrinos have the property, however, of being able to pass, unaffected, through great masses of matter. A hundred billion neutrinos, moving at the speed of light, pass through each square centimeter of the Earth's cross-section each second — as though the Earth were not there. Only very occasionally does a neutrino interact with one of the particles of the planet.

After 1956, strenuous efforts were made to detect these Sunborn and star-born neutrinos by means of these occasional interactions. Reines, for instance, set up a large apparatus, designed to detect such neutrinos, in a South African gold mine two miles deep in the Earth. It may seem strange to attempt to study the heavens from a hole in the ground, but neutrinos can reach that hole (or any other portion of the Earth down to its very center) without trouble, whereas no other detectable particle can. Finally, in 1965, after observations lasting half a year, Reines reported the detection of seven neutrinos (seven!).

In 1968, a still more elaborate neutrino-trap was set up by Raymond Davis, Jr. in a deep mine in South Dakota. Solar neutrinos were indeed detected but in only about one-sixth to one-third the quantity which had been set as the minimum to be expected by any of the schemes currently used to describe what went on in the core of the Sun. Either our notions about the inner workings of stars must be fundamentally revised, or some point is being overlooked that will turn out to be "obvious" once it is explained, or the method of neutrino detection just isn't reliable enough.

In any case, all this is the first birth pang of "neutrino astronomy." If, in times to come, the methods of detection become more efficient, much of value may be learned. Neutrinos come directly from stellar cores and by analyzing the distribution of energies among them, the temperature and other properties of the core may be determined directly, rather than deduced more or less uncertainly by indirect means.

Cosmic Rays

Much more spectacular is the bombardment of the Earth by particles possessing mass.

As early as 1900, nuclear physicists were studying the manner in which energetic radiation from radioactive atoms knocked electrons out of atoms in the atmosphere. What was left of the atoms carried a positive electric charge and these charged atom-fragments were called "positive ions."

Gradually, it became clear that no matter how carefully physicists shielded a sample of air, enclosing it in lead boxes that were supposedly impervious to radiation, ions continued to be formed within the sample at a slow rate. Apparently, radiation still more energetic than had yet been observed was penetrating a thickness of lead that would have been an adequate shield under ordinary circumstances.

Physicists, generally, took it for granted that this particularly energetic radiation was coming from the soil. It was in the soil, after all, that radioactive susbtances were to be found. To settle the matter, the Austrian physicist Victor Franz Hess (1883-1964) undertook a series of balloon ascensions beginning in 1911. The plan was to test for the presence of this ionizing radiation some miles above the Earth's surface. The thickness of air between the surface and the balloon ought to suffice to absorb at least some of the energetic radiation, and the rate of ion-formation in the balloon would then prove to be less than that on the surface.

Precisely the reverse was observed. The rate of ionization increased, and the higher the balloon rose, the greater the increase. Clearly, the origin of the radiation was not below the balloon on the Earth's surface, but above it. Hess called it "high-altitude radiation." Over the next decade or so, Hess's results were confirmed over and over again, and it became quite plain that the radiation was striking Earth everywhere and must have its source somewhere out in space. Since the radiation reached Earth from the outside Universe (or cosmos), the American physicist Robert Andrews Millikan (1868-1953) called it "cosmic rays" in 1925, and that name stuck.

The question next arose as to just exactly what the cosmic

rays might be. One of two alternatives seemed most likely: they were either extremely energetic photons, shorter in wavelength and therefore more energetic than any ever observed previously, or they were massive particles moving at extremely high velocity and deriving their unprecedented energies from the combination of great mass and high velocity.

All the massive particles known in the 1920's carried an electric charge and it was assumed that would be true of the cosmic rays as well if they fell into this category. It followed from this that if the cosmic rays approached the Earth from all directions equally, they ought, nevertheless, to strike the polar regions with greater frequencies than they struck the Tropics.

This arose from the fact that the Earth behaved like a magnet, with its north and south magnetic poles in the polar regions, and with magnetic lines of force curving outward from north to south, spreading most widely apart in the Tropics and coming most closely together in the polar regions. Any charged particle entering the Earth's magnetic field would be deflected northward or southward according to an interaction that was well-known to the physicists of the time. Some particularly energetic cosmic rays might smash right through the magnetic field by main force and

The Earth as magnet

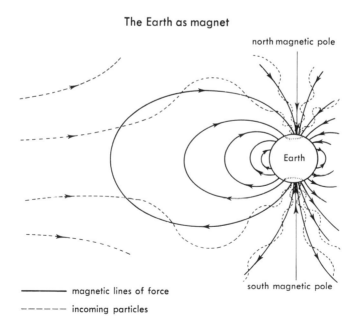

magnetic lines of force

incoming particles

land in the Tropics (if aimed there), but on the whole the incidence of cosmic rays would increase steadily as one traveled north or south from the Equator.

Photons, on the other hand, being uncharged, would remain essentially unaffected by the Earth's magnetic field. If cosmic rays were photons and approached the Earth from all directions equally, they would, in that case, strike all portions of the Earth's surface equally.

Compton, beginning in 1930, carried out extensive studies of cosmic rays in different portions of the world and found that there was indeed an increase in intensity with latitude, precisely as one would expect if the rays consisted of charged particles.

There was still the question as to whether the particles were positively charged or negatively charged. The Italian physicist Bruno Rossi (1905-   ) pointed out in 1930 that the nature of the deflection of cosmic rays would vary with the nature of the charge. Positively charged particles would be deflected in such a way that more would seem to be approaching the observer from the west than from the east, while the reverse would be true for negatively charged particles. By 1935, the decision was clear: cosmic rays were made up of positively charged particles.

This is true at least for the radiation in the form in which it exists before it hits the Earth's atmosphere ("primary radiation"). Once it enters the atmosphere, it strikes atoms, producing a variety of kinds of particles of lesser, but still very high, energy ("secondary radiation"). It is naturally with the primary radiation that astronomers are chiefly concerned.

To reach the primary radiation one must rise into the stratosphere at least, leaving most of the interfering, interacting air molecules beneath. In the late 1930's balloon ascensions began to be made for the purpose, making use of new plastic materials, lighter and more leakproof than the products available earlier. Altitudes as high as thirteen miles were achieved, and at that height 97 percent of the atmosphere had been surmounted.

The primary radiation, it turned out, consisted chiefly of protons (hydrogen-1 nuclei) moving at nearly the speed of light. By the 1960's it was clear that protons made up about 90 percent of all the energetic particles of the primary radiation. An additional 9 percent were made up of helium-4 nuclei, and the remainder of nuclei of still heavier atoms, up to and including even those past uranium — the most complicated atoms that occur in quantity on earth. The first nuclei more massive than those of uranium were detected among the cosmic-ray particles in 1967.

In short, cosmic rays were made up of the general matter of interstellar space, set into violent motion, and possessing almost unimaginable concentrations of energy.

In the last few years, physicists have managed to build costly, enormous, and intricate "particle accelerators" that can endow single particles with as much as thirty billion electron-volts ("30 Bev") of energy. The size of this energy concentration may be judged from the fact that the energy of an ordinary photon of light might be something like two electron volts ("2 ev").

Man, then, can produce protons with energies up to fifteen billion times that of a light photon, and such energies do indeed represent an approach to the level of energies in cosmic rays (which is the reason why one giant particle accelerator was given the name "Cosmotron" by its designers).

Nevertheless, only the lowermost levels of cosmic ray energies are reached even by man's proudest creations. Many cosmic ray particles strike the Earth's atmosphere with energies of far more than 30 Bev; some reach us with energies of 1,000,000 Bev and more!

Cosmic Ray Sources

But where do the cosmic rays come from? And what gives some of them such incredibly tremendous energies?

A logical origin might be the Sun, but that was almost immediately eliminated as a main source since cosmic rays approach the Earth from all directions equally, from the direction opposite that of the Sun just as frequently as from the direction of the Sun. Granted that cosmic rays, if originating in the Sun, would be deflected by the Earth's magnetic field, and that some would even be pushed around to the rear, it is quite unthinkable that the end result of any such deflection would be to smear the cosmic rays so completely evenly all about the Earth. The source had to be, then, from somewhere outside the Solar system. The even smear of cosmic ray influx made it impossible, however, to pin the phenomenon down to any particular objects in the Universe.

And yet the Sun was not to be counted out completely.

The Sun's radiational system is not perfectly even and smooth. The Solar surface breaks into "sunspots," regions of comparatively low temperature which, therefore, show up black against the hotter and brighter surroundings. These sunspots are accompanied by magnetic fields, and the energy store in these fields can manifest itself in violent manner.

One of these manifestations is that of the "Solar flare," the sudden brightening of an irregular area near a sunspot. The first report of a Solar flare was made in 1859 by the English astronomer Richard Christopher Carrington (1826–1875). He thought the sudden brightening on the Solar disc was caused by the fall of a large meteor (in line with the suggestions Helmholtz was making at the time). Almost immediately after that observation, however, disorders were reported in the behavior of compasses and the aurora borealis in polar regions grew particularly brilliant.

Since then the association of Solar flares with such "magnetic storms" has been unmistakable. Not all flares give rise to them, of course, only those that are more or less central on the Sun's face and that are therefore aimed directly at us. It seems clear that the great energies associated with flares suffice to eject quantities of subatomic particles out into space. Since flares and other energetic phenomena are always taking place here and there on the Sun's surface, we might picture the Sun as surrounded by a cloud of energetic charged particles shooting outward from it in all directions. This is the "Solar wind."

By 1958, rocket experiments had proved the actual existence of this Solar wind. The velocity with which particles stream outward from the Sun can be as high as 450 miles a second, and such particles carry a surprising amount of kinetic energy. Until the last decade, for instance, it had been thought that comets' tails streamed away from the Sun because of the pressure of Solar radiation against the tiny particles making up the tails. Apparently, this is not so; it is the force of the Solar wind that does it.

Naturally, the charged particles in the Solar wind, on approaching the Earth, would have to interact with the planet's magnetic field. As predicted by the Greek amateur scientist (and now professional physicist) Nicholas Christofilos in 1957, the charged particles making up the wind would be deflected by the magnetic lines of force, spiralling about them from north magnetic pole to south magnetic pole and back, contributing to the formation of a doughnut-shaped region of dense charge about the Earth, well above its atmosphere. This was not taken seriously at first (partly because Christofilos was an amateur), but in 1958, rocket observations made under the direction of the American

physicist James Alfred Van Allen (1914–   ) demonstrated
the actual existence of such regions. They came to be known as
the "Van Allen belts" at first and then, later, as the "magnetosphere."

The limits of the magnetosphere are shaped into a stream-
lined teardrop by the Solar wind. In the direction of the Sun, the
magnetosphere is bluntly rounded, but it curves into a long tail
on the side away from the Sun.

The presence of a Solar flare pointed in our direction produces a local strengthening of the Solar wind, a kind of storm-
blast of charged particles bearing down on us and filling the magnetosphere to overflowing. The particles pour down on the polar
region of the Earth, particularly, giving rise to the strengthened
aurora and introducing such irregularities into the magnetic field
as to send compasses skittering. More important in a practical
sense, nowadays, a Solar flare alters the properties of those sections of the upper atmosphere which normally contain a high
concentration of electron charge in the form of ions. (That section of the atmosphere is therefore called the "ionosphere"). As
a result, magnetic storms upset the workings of radio and, indeed,
any device that interacts with the radiation of the ionosphere.

Naturally, the energy of the particles released into the Solar
wind varies according to the size and force of the flare. What if
a really large flare burst out? One took place in 1942 and it
was quickly followed by a temporary increase in the cosmic ray
influx. This has been observed a number of times since, and it
became clear that the Sun *can* serve as a source of cosmic rays,
at least occasionally. These solar cosmic rays are "soft"; that is,
they are relatively unenergetic, with energies ranging from 0.5 to
2 Bev, but the principle remains.

What if stars generally produce cosmic rays as a result of
flares or other phenomena? Perhaps in passing across the vast
interstellar distances, local magnetic fields would deflect the cosmic
ray particles over and over again so that all signs of the original
direction of travel would disappear. Such local magnetic fields do
indeed exist. The Sun has a strong magnetic field, for instance, one
that is concentrated at the sunspots — as was shown as long
ago as 1907 by the American astronomer, George Ellery Hale
(1868–1938). Similar fields were detected in certain other stars
in 1947 by the American astronomer, Horace Welcome Babcock
(1912–   ).

In the end, then, thanks to such fields, the motions of cosmic ray particles would be randomized to the point where they would seem to come from all directions equally rather than in highest concentrations from the plane of the Milky Way, where most of the stars are.[3]

This explanation is not quite sufficient. If all stars contributed equally to cosmic ray production, the Sun, being much the nearest, would drown out all the rest, as it does with respect to light production. In that case, the cosmic ray influx would appear to be heavily weighted in the direction of the Sun — and it is not.

It follows, then, that some stars are much richer producers of cosmic ray particles than the Sun is. There are certain variable stars, for instance, whose source of variation is the periodic production of large flares. Possibly, these "flare stars" are rich sources. And then, too, there are the supernovae.

The cosmic ray influx from such specialized members of the Galaxy may completely drown out the piffling production from ordinary stars, even from our own nearby Sun.

But that still leaves the problem of the energies of the cosmic ray particles. If the Sun can produce particles with 1 BEV of energy, it is not surprising that a supernova could produce them with much larger energies — but billions of BEV? No known nuclear reaction in even the hottest and most ferocious supernovae could be expected to produce particles as energetic as many of those in cosmic rays.

But do all the particles actually have to be formed at those energies to begin with? In 1951, the Italian-American physicist Enrico Fermi (1901-1954) suggested an alternate possibility. Suppose that cosmic ray particles were produced at quite moderate energies of a few BEV and suppose that the effect of the Galaxy's magnetic field was to accelerate such particles and increase their energies.

The process envisioned was similar to that of man-made cyclotrons, devices which whirled charged particles round and round under the influence of a magnetic field, pumping additional energy into them at every cycle. As the particles gain energy, they are deflected less and less by the magnetic field until finally they can no longer remain within the confines of the cyclotron. They shoot out at high energies.

[3] Not only would most stars be expected to have an associated magnetic field that would serve to deflect any cosmic ray particles passing by, but the Galaxy generally is now thought to have a weak magnetic field of its own — one that may serve as a factor in maintaining the existence of the arms of spiral galaxies.

The magnetic intensities of man-made magnets are much greater than those of the Galaxy's magnetic field, but the latter extend over many thousands of light-years. The cosmic ray particles are accelerated only very slowly as they travel, but in eons of time they reach high energies indeed.

At any point in their travels, such cosmic ray particles may happen to smash into some obstacle such as our own planet. We would therefore be subjected to a wide spectrum of particle energies, for the energy of any given particle would depend to a large extent on how long it had been traveling in space before colliding with us. The longer the time of previous travel, the higher the energy at the time of collision.

As a cosmic ray particle gains energy, however, it is deflected less and less under the influence of the magnetic field until it is finally traveling in so nearly a straight line that even the vast width of the Galaxy is insufficient to hold it. By the time the energy of the particle reaches the neighborhood of a hundred million BEV, it goes shooting out of the Galactic cyclotron, so to speak.

It would be expected then that if our Galaxy were the only source of cosmic ray particles, no energies higher than a hundred million BEV would be detected. Nevertheless, higher energies *are* occasionally detected, up to ten billion BEV at least. It can only be assumed that such super-energetic particles must originate in other galaxies with stronger magnetic fields than our own, galaxies that have exploded, imploded, or, in general, experienced catastrophic changes beyond that of the ordinary supernova. Such galactic catastrophes would liberate vast energies and floods of cosmic rays in the billions of BEV.

These, after finally being hurled out of their parent galaxies without having had the ill-fortune of having collided with any piece of matter, cross intergalactic space, happen to pass through our own Galaxy and — strike us.

17

Energetic Photons

The Electromagnetic Spectrum

But let us now return to the photon. Until 1800, the only photons known to man were those of visible light, which he could sense directly. The wavelength of such light varies from 0.000076 centimeters at the red end of the visible spectrum to half that value, or 0.000038 centimeters at the violet end. The energy of light photons is inversely proportional to the wavelength of the light. If extreme violet light has half the wavelength of extreme red light, then those violet light photons have twice the energy content of the red light photons. The energy content of the photons of visible light varies from 1.5 electron volts (1.5 EV) at the extreme red end of the spectrum to 3.0 EV at the extreme violet end.

In the opening years of the nineteenth century, infrared and ultraviolet radiation were discovered (see page 64). The energy content of infrared photons was, naturally, less than 1.5 EV, while that of ultraviolet photons was more than 3.0 EV.

How far the infrared region of the spectrum might extend in the direction of lower and lower energies and how far the ultraviolet region might extend toward higher and higher energies was unknown through most of the nineteenth century.

In 1861, however, the Scottish physicist James Clerk Maxwell (1831–1879) evolved an overall theory of electricity and magnetism that showed the close and, indeed, inseparable relationship of these two types of energy. (One can only speak of an "electromagnetic field" as a consequence, taking the two types of energy together.) He also showed that periodic variations in the intensity of such a field would produce a wave-form that would recede from the source of variation at the speed of light. Indeed, light itself was considered a form of such "electromagnetic radiation."

Since the electromagnetic field can vary with any period, the electromagnetic radiation can have any wavelength. There should, therefore, exist electromagnetic radiations with wavelengths far longer than even the infrared, and others with wavelengths far shorter than even the ultraviolet.

It did not take long for the prediction to be verified. In 1888, the German physicist Heinrich Rudolf Hertz (1857–1894) produced electromagnetic waves with enormous wavelengths. Such radiation (called, at first, "Hertzian waves") came to be used for radiotelegraphic communication—that is, communication not by electric currents along wires as in ordinary telegraphs, but by waves radiating (hence the "radio-") through space. Naturally, one might expect to have such radiation called radiotelegraphic waves for this reason, but the shortened form "radio waves" came into fashion.

In 1895, the German physicist Wilhelm Konrad Roentgen (1845–1923) demonstrated the existence of a form of radiation that turned out to be electromagnetic in nature and to have extremely short wavelengths. He called them "X-rays" as a confession of ignorance as to their nature, and the name stuck, even after the ignorance had vanished.

The three varieties of radiation from radioactive substances (discovered first in 1896 by Becquerel) were named by Rutherford after the first three letters of the Greek alphabet: "alpha rays," "beta rays" and "gamma rays." Of these, the gamma rays proved to be electromagnetic in nature; a form of radiation with wavelengths even shorter than those of X-rays.

By the beginning of the twentieth century, then, physicists found themselves in possession of an enormous "electromagnetic spectrum" stretching over some sixty octaves—that is, with wavelengths doubling sixty times as one progressed from the shortest to the longest. The longest waves were therefore 2^{60} or about

1,000,000,000,000,000,000 (a billion billion) times as long as the shortest. Of this vast range, visible light covered only a single octave.

The electromagnetic spectrum is continuous, and there are no gaps between one form of radiation and another. The boundaries man sets are purely arbitrary, depending on his ability to sense one small portion directly, and on the accidents of discovery outside that portion. Ordinarily, these arbitrary boundaries are described in terms of wavelength, or of frequency (the number of wavelengths produced per second). I will, instead, describe them here in terms of the energy content of the photons making them up, a value that is directly proportional to the frequency.

The electromagnetic radiations of longest wavelengths (and, therefore, made up of photons of least energy) are the radio waves. At the broadest extent, they contain photons with energy contents of 0.001 EV and less. This turns out to be an uncomfortably large range, and it is often broken up into three regions: long radio waves, short radio waves, and very short radio waves. The last are now frequently referred to as "microwaves." The energy contents of the photons would be as follows:

Long radio waves—zero to 0.00000001 ev
Short radio waves—0.00000001 to 0.00001 ev
Microwaves—0.00001 to 0.001 ev

The infrared region can, in turn, be divided into the far infrared, the middle infrared, and the near infrared, as we move along the scale of shortening wavelength and of increasingly energetic photons.

Far infrared—0.001 to 0.03 ev
Middle infrared—0.03 to 0.3 ev
Near infrared—0.3 to 1.5 ev

The visible region is, as stated before, in the range of 1.5 to 3.0 EV. By color, the energies might be listed as (on the average):

Red—1.6 ev
Orange—1.8 ev
Yellow—2.0 ev
Green—2.2 ev
Blue—2.4 ev
Violet—2.7 ev

Electromagnetic radiations made up of photons with energies higher than those of visible light include, in order, the near ultraviolet, the far ultraviolet, X-rays, and gamma rays:

Near ultraviolet—3 to 6 ev
Far ultraviolet—6 to 100 ev
X-rays—100 to 100,000 ev
Gamma rays—100,000 ev and up.

X-ray Stars

Naturally, the question arises as to how far over the whole electromagnetic spectrum the observed spectra of the Sun and stars extend. It is certain that the Solar spectrum is not confined to the visible octave, for both infrared and ultraviolet radiation were first discovered in the Solar spectrum.

But there is a sharp limit to how far beyond the visible edges of the Solar spectrum investigations can be carried on, at least on the Earth's surface. The atmosphere, while transparent in the visible light region, is quite opaque for almost all other sections of the electromagnetic spectrum. Though the Sun might well be rich in radiation in the far ultraviolet and far infrared, such radiation would nevertheless not reach us under our blanket of miles of air, and until well into the twentieth century, nothing much could be done about that.

By the mid-twentieth century, however, man's technology had made the atmosphere no longer the impenetrable barrier it had once been. Airplanes could climb into the stratosphere and remain there for hours; balloons could rise even higher and remain there for days; rockets and satellites could rise beyond the atmosphere altogether and remain there for weeks and months, even years.

Astronomic observations made from the stratosphere and beyond could take advantage of the entire energy range of photons, and not merely of those few varieties that could penetrate, relatively uninterfered with, to the bottom of the ocean of air under which we move and live.

In 1964, for instance, a balloon-borne telescope studied the infrared portion of the light reflected by the planet Venus. The absorption bands present in the infrared spectrum of the planet made it quite plain that its clouds contained ice crystals and were therefore probably made up of water. It is difficult or impossible to reach such a conclusion, unmistakably, from the surface of the Earth. From the surface, the necessary portion of the spectrum remained hidden, and any effect of the water content of Venus' atmosphere on that portion of its spectrum visible from Earth's surface was bound to be more or less masked by the water vapor present in Earth's own atmosphere.

Again, from balloons, photographs of the Sun could be taken that were much sharper than any that could be taken from the Earth's surface. The Solar spectrum could be carried far out into the ultraviolet and, especially after rockets came into use, thousands of absorption lines could be recorded and measured that would have been forever invisible from down here.

And it is not just the Sun that benefits from this extension of the spectral reach. In 1968, observations from satellites in orbit beyond Earth's atmosphere showed that ultraviolet radiation from the nuclei of several galaxies, including Andromeda, was several times as intense as had been expected.

Nor need we stop at the ultra-violet. Even as early as 1949, the Sun's spectrum was discovered to spread into the X-ray region to a much greater extent than would have been thought possible. In general, the higher the temperature of a body, the greater the quantity of high-energy photons it emits. This was, in effect, shown by Wien (see page 120) in the days before the existence of photons was suspected. And the Sun's surface temperature was certainly not high enough for X-rays.

The Sun's surface temperature of 6000° C. placed the peak of its radiation in the visible light region, and the photons it emitted most profusely from the surface would have an energy content of several electron-volts. To emit large quantities of X-ray photons with energies of hundreds and even thousands of electron-volts meant that much higher temperatures had to be available. Such temperatures did not occur on the untroubled surface of the Sun, but they did occur in the Sun's "corona" (its rarefied outer atmosphere). Here a thin scattering of particles was whipped into extremely rapid motion, probably by shock waves from the lower atmosphere. The energies imparted to these particles were equivalent to temperatures of 500,000° C., at least. When a Solar flare erupted the temperature of the corona above the flare would rise upward into the millions of degrees.

To be sure, although individual particles of the corona were exceedingly energetic, the total heat of the corona was small because the total number of particles was comparatively small. The heat of the corona carries no danger to the Earth, therefore. (Indeed, there is some reason to think that the outermost wisps of corona extend outward from the Sun to beyond the Earth's orbit so that, in a sense, the Earth actually revolves within the Sun's atmosphere—but without being sensibly affected thereby.)

It is the corona and the flares that are the source of Solar X-rays. And it is the high temperature-equivalences of these regions that strip so many electrons from atoms as to produce utterly novel spectral lines—lines that in earlier years were interpreted as representing a new element called coronium (see page 131).

The search for energetic photons from sources other than the Sun was also undertaken. Energetic photons would have enormous advantages over the energetic particles of cosmic rays in some ways. Photons, being uncharged, would be unaffected by magnetic fields and would travel in undeflected trajectories. Photon source could, therefore, be identified; at least, the direction from which they approached could be determined. Thus, regions particularly rich in ultraviolet radiation were uncovered in 1956 and afterward, and it was not difficult to tell that the most prominent of these were centered in the constellations Orion and Virgo.

In Orion, the energetic radiation seemed associated with the luminous nebulae surrounding particularly hot stars. In a way, these nebulae seemed to be analogous to vastly extended coronas, heated by the superabundant energy of the stars within, in the same manner that the solar corona is heated by the Sun.

Again, X-ray photons produced some surprises. The Sun radiated X-rays, true, but only to a minor extent, in the absence of flares. Solar X-rays were detected only because the Sun was so close. If other stars did no better, they could not be expected to register their X-rays over vast interstellar distances.

Nevertheless, a team of investigators, including Bruno Rossi, was interested in finding out whether the Solar X-rays might be reflected from the Moon (the surface of which, unlike that of the Earth, was *not* protected by an atmosphere). Special rockets sent up in 1962, with instruments designed to detect X-rays, managed to detect them, not from the Moon, but from the general direction of the Galactic center.

The next year rockets were sent up by a group under the leadership of the American astronomer Herbert Friedman (1916–). It was their intention to scan the sky for X-ray sources and pinpoint their locations as closely as possible. Some thirty regions showing X-ray activity were detected in this way over the next couple of years, concentrated in the direction of the galactic center. There may be as many as 1250 in the entire Galaxy. The strongest of these, perhaps only 100 light-years away, was found in the constellation Scorpio. This source, now referred to as Sco XR-1, may have been the one that tripped the detecting devices in that first 1962 rocket flight.

The second strongest source, one-eighth the strength of the Scorpio source, seemed to come from the Crab Nebula.[1] (Tau XR-1.) Of course, the comparison may be misleading. Sco XR-1 is the stronger because it is so much the closer. At equal distances, the Crab Nebula would be perhaps a hundred times as intense an X-ray source as Sco XR-1.

The nature of the objects radiating so strongly in X-rays that they could be detected across many light-years of space was a most intriguing question. Again, temperatures in the millions of degrees had to be evolved, but the total energy would have to be many times greater than that of the Sun's corona to produce so much. One might almost expect that a star's core would have to be exposed. Is it the result of a supernova explosion? The Crab Nebula is the remnant of such an explosion a thousand years ago, and Sco XR-1 may be the remnant of a 50,000-year-old supernova. Yet even that may not be enough.

Such temperatures might arise out of catastrophic contractions still greater than those that would be involved in the conversion of supernovae into ordinary white dwarfs. The degenerate matter that composes a white dwarf is made up of protons, neutrons, and electrons. Of these, the electrons are the crucial components. They are less compact than are the protons and neutrons and therefore resist compression more than the latter do. As long as the mass of the white dwarf is no greater than 1.4 times that of the Sun (see page 168), the electrons will manage to keep the star expanded to planetary volumes, at least, even against the enormous compressing force of its extremely compacted gravitational field. If the white dwarf possesses a mass greater than 1.4 times that of the Sun, then even the electrons could not resist the compressive force of the still more intense gravitational field.

An ordinary star with a mass more than 1.4 times that of the

---

[1] The Crab Nebula added to its remarkable properties in another way in 1964. In that year, cosmic-ray counts taken by balloon showed higher values when the Crab Nebula was in view of the detectors. If this is not coincidence, then the Crab Nebula is the only known specific cosmic-ray source other than the Sun itself.

Sun would, in undergoing a supernova explosion, lose a large part of its mass, and what is left may usually be expected to be below the crucial mark for white dwarf formation. But what if it is not?

In that case, compression does not stop at the white dwarf stage. There is further compression; the electrons melt into the protons, forming neutrons, and the whole comes together into a mass of neutrons-in-contact. Such a "neutron star" would contain all the mass of a couple of Suns in a sphere not more than a dozen miles in diameter. It would be made up of neutronium, or of Gamow's ylem. It would be a tiny fragment of the substance thought by those favoring the big bang theory to have made up the cosmic egg. The very existence of such neutron stars would be a small point in favor of the big bang theory for that very reason.

Theorists suggest that a neutron star, for a period after its formation, would have an equal temperature throughout its structure, a temperature of about 10,000,000° C. It would consequently release a furious flood of X-rays and would thus be a most effective X-ray source. A neutron star would therefore be an "X-ray star" as well.

The question of the existence of a neutron star must be put, if possible, to the test of observation. If the X-ray source is really a neutron star, then the X-rays are emerging from a point in space. No possible detecting device we could build would make a body a mere dozen miles in diameter anything more than a point when it is at a distance of thousands of light-years from us as the Crab Nebula is. On the other hand, if the X-rays were emerging from a comparatively large area of space, then the source is likely to be a turbulent region of gas and dust, and the case for a neutron star is weakened. (It is not entirely eliminated, however, since the area of gas and dust might still surround a neutron star, with both radiating X-rays.)

It was possible to differentiate between a point source and an area source of X-rays as a result of an interesting astronomic accident. The Crab Nebula is situated in such a position that it is periodically covered (or "occulted") by the Moon. As the Moon passes before it, the X-ray source might be cut off instantaneously, indicating it to be a point, or it might be cut off slowly and gradually, indicating an area.

In 1964 an occultation of the Crab Nebula by the Moon was scheduled to take place. If astronomers missed their chance, they would not have another for eight years. A rocket was sent up in time by Friedman's group, and its instruments worked. The X-ray flux was found to fade out gradually, and the source was shown to be an area about a light-year in diameter at the center of the Crab Nebula.

For a while (but, as it turned out, only for a while) the possibility of neutron stars had to be dismissed, especially when Sco XR-1 was identified with a 13th magnitude star in 1966. After all, it didn't seem likely that a neutron star could be seen at astronomic distances.

Nevertheless, if the neutron-star notion appeared to have gone glimmering, X-ray sources offered a great many other puzzles to interested astronomers. If Sco XR-1, for instance, were not a neutron star, what would account for its flood of X-rays, which it radiated with a thousand times the intensity of the visible light it radiated.

Then, too, it was soon discovered that X-ray sources could be markedly variable. A strong X-ray source in Cygnus, Cyg XR-1, decreased considerably in X-ray brightness in the course of a single year. Another one, Cent XR-2 (in Centaurus), which was apparently non-existent in October 1965, suddenly flared and increased in intensity till it was the brightest X-ray source in the sky in April 1967 and then faded to a tenth its maximum in September 1967. It might almost be considered an "X-ray nova."

Finally, X-ray sources were located outside the Milky Way Galaxy. In 1968, one was located in the Large Magellanic Cloud, which may be up to 150 times as intense as the Crab Nebula source. What's more, at least two of the X-ray sources turned out to be associated with other galaxies: Cygnus A and M-87. These were the first "X-ray galaxies" and there may be as many as 10,000 detectable X-ray galaxies in the Universe.

Antimatter

Gamma-ray photons, more energetic even than X-ray photons, have also been detected in outer space, by means of satellite-borne instruments. In 1961, twenty-two such photons were detected by instruments on the Explorer XI satellite. The direction from which these photons arrived did not seem to be confined particularly to the plane of the Milky Way, and the conclusion was that they came from other galaxies. In 1965, the first point-source of gamma rays was detected, so that now we will be dealing with "gamma-ray stars."

Naturally, one must try to think of a source for these gamma rays. They seem to be rare events so that one need not imagine a massive source but can postulate isolated subatomic events. A collision of a highly energetic cosmic ray particle with any atomic nucleus would set off a train of consequences that would include the production of gamma rays. The gamma rays we detect may, then, mark the burial spot of dead cosmic ray particles.

Another possibility is more dramatic and requires a bit of prologue. By 1932, three types of subatomic particles were known. These were the proton (massive, positively charged), the neutron (massive, uncharged), and the electron (light, negatively charged). Of these, atoms are composed; and of atoms, matter is composed.

In 1930, however, the English physicist Paul Adrien Maurice Dirac (1902–   ) had suggested, out of purely theoretical considerations, that for each type of particle an "antiparticle" ought to exist. These antiparticles would be marked by key properties exactly opposed to those of the corresponding particles.

Thus, equivalent to the negatively charged electron would be an "antielectron" similar in all respects except for the possession of a positive charge equal in size to the electron's negative charge. Similarly, balancing the positively charged proton would be a negatively charged "antiproton." The neutron is uncharged, but it has a magnetic field oriented in a certain direction. Balancing it there would be an uncharged "antineutron" with a magnetic field oriented in the opposite direction.

It seemed a rather far-out suggestion at first but in 1932, the American physicist Carl David Anderson (1905–   ), in the course of his studies of cosmic rays, discovered the antielectron. In reference to its positive charge, he called it the "positron," by which name it is most commonly known to this day although antielectron is the more appropriate. The antiproton and antineutron were detected in 1956.

The antiparticles possess all the properties of the particles (aside from their mirror-image reversals) and can do anything particles can do. If protons and neutrons can combine to form atomic nuclei, there is no reason why antiprotons and antineutrons cannot combine to form atomic "antinuclei." Indeed, in 1965, antiprotons and antineutrons were combined at Brookhaven to form a combination made up of one of each. Since a proton plus a neutron make up the nucleus of hydrogen-2, or deuterium, the proton-neutron combination is called a "deuteron." The combination of an antiproton and an antineutron is therefore an "antideuteron."

Then, just as an atomic nucleus can surround itself by electrons to form a neutral atom, so an atomic antinucleus can surround itself by antielectrons (positrons) to form a neutral "antiatom." And as atoms make up matter, so antiatoms make up "antimatter."

The difficulty of constructing such antimatter in the laboratory rests in the evanescent nature of the antiparticles. Left to themselves the antiparticles would be as stable as the particles to which they correspond, but they are not left to themselves. When an antiparticle is formed, it comes into existence in a Universe of ordinary particles; it is isolated in a vast ocean of its opposites.

If an antielectron is formed, it is only a matter of time (a millionth of a second or less) before it meets and collides with an electron. The result is "mutual annihilation." The charges cancel, and the total mass of the pair is converted into energy in the form of photons. The same is true of the collision of a proton and antiproton.

The mutual annihilation of particles and antiparticles offers a new and unprecedentedly powerful source of energy. The most energetic ordinary nuclear reactions, such as those which fuse hydrogen to iron in stellar cores, involve the loss of only some 1 percent of the total mass. In the mutual annihilation of matter and antimatter, *all* the mass is converted into energy. Mass for mass, then, mutual annihilation produces a hundred times as much energy as nuclear fusion.

The reverse process can also take place. A gamma ray of appropriately high energy can be converted into an electron-antielectron pair. A still more energetic gamma ray can be converted into a proton-antiproton pair. (This last process, long predicted by theory, was finally observed in 1965.)

But now a difficulty arises. Careful observation makes it appear as though there is a "law of conservation of electric charge." Negative electric charge can neither be created nor destroyed by itself; nor can positive electric charge. What can happen is that equal quantities of positive and negative electric charge can undergo mutual annihilation and conversion to gamma ray photons, in which case there is no destruction of *net* electric charge. Similarly, positive and negative electric charge can be created in equal quantities out of gamma ray photons so that there is no creation of *net* electric charge.

But in that case, how is it that we are surrounded by a Universe of matter, without any appreciable sign of antimatter? After all, any process that creates particles ought to create antiparticles as well, and in equal quantities.

Still, how do we know that we inhabit a Universe made up of matter only?

We can be sure that the Earth itself is exclusively matter without any significant admixture of antimatter, for if any antimatter were present, it would interact with matter at once and disappear in a blaze of gamma rays. The Moon, too, is matter, if only because man-made rockets have landed on it without producing a colossal explosion.

Meteorites are composed of matter[2] and the Sun is matter, if only because the Solar wind is composed of particles rather than antiparticles. Consequently, it is safe to conclude that the Solar system is matter.

Since the cosmic rays are composed almost exclusively of particles and not antiparticles, we can even say that the Galaxy (and perhaps nearby outer galaxies as well) is matter.

But in that case, how can one explain the nonexistence of antimatter?

Consider the steady-state model of the Universe. If there is a continuous creation of hydrogen atoms (protons and electrons), why is there not an equivalent creation of hydrogen antiatoms (antiprotons and antielectrons)? If there is a continuous creation of neutrons, which then break down to protons and electrons, why is there not a continuous creation of equal numbers of antineutrons that break down to antiprotons and antielectrons?

We know of no reason to explain the preference of one over the other, and all the evidence gathered by physicists so far would lead us to believe that no preference can be shown.

One possibility, of course, is that both matter and antimatter are indeed continuously created but that some mechanism balances the process in such a way that they are created in different places. An atom may come into existence *here* and be balanced by an antiatom coming simultaneously into existence *there*. In that case, perhaps, there are both matter-galaxies and antimatter-

---

[2] Occasionally, the existence of large meteorite strikes without much in the way of meteoric matter at the site of the strike leads to speculation concerning the possible landing, once in a long while, of an antimatter meteorite.

galaxies; or, to put it more compactly, both galaxies and anti-galaxies.

If this is so, could we distinguish between a galaxy and an antigalaxy?

Not, apparently, through the light they emit. A photon is its own antiparticle, so that both matter and antimatter, inter-acting among themselves, produce photons of identical nature. The light from an antigalaxy is just like the light from a galaxy.

The question of gravitational effects is less certain. There are speculations that while matter undergoes a mutual gravitational attraction, and antimatter the same, the gravitational interaction beween matter and antimatter is a repulsion. No gravitational repulsion has ever been observed, but then antimatter has never been studied in quantities sufficient to produce a perceptible grav-itational field, so the matter must be considered unsettled.

If such gravitational repulsion exists, it should be encountered between galaxies and antigalaxies. None has been, which might indicate that no antigalaxies exist in the Universe. It might also mean that the effects would be small indeed at intergalactic dis-tances as seen across intergalactic gaps of space, and that such effects exist but have not yet been observed or been interpreted properly.

Neutrinos offer a more definite hope. Galaxies release floods of neutrinos and antigalaxies floods of antineutrinos. If areas of the sky can be pinpointed as rich sources of antineutrinos, anti-galaxies may be located. However, neutrinos are extremely diffi-cult to detect and astronomic art is not yet at the point where such a feat is practical.

Again, antigalaxies ought to produce cosmic-ray antiparticles. Those likely to reach us would be very few in number, but some might and those that do would be useful. Antiparticles of billion Bev energies and more (the energies required for cosmic rays to shake loose of their galaxies and streak across intergalactic space toward us) would be but little affected by galactic magnetic fields. The direction of their arrival might serve to pinpoint antigalaxies.

It is important to remember, however, that galaxies are not in isolation, but that innumerable galaxies exist and that some may interact despite the general expansion of the Universe. A galaxy and antigalaxy might belong to the same cluster, for instance, and might approach. If they approach closely enough for their dust and gas to begin intermingling at the fringes, vast quantities of energy would be emitted at once. There are, in fact, cases of vast energy-release in the depths of galactic space that might just possibly indicate such matter-antimatter mutual annihilation. I will return to this subject later in the book.

It may be that galaxies and antigalaxies are actually protected from massive annihilation by such reactions at the fringes. An analogy can be drawn with a drop of water falling on a hot stove. The water does not evaporate immediately in an explosion of boiling. Instead, it hops and skips on the red-hot surface for a surprisingly long time. The reason for that is that the portion of the drop initially touching the hot surface vaporizes, and a cloud of steam pushes the drop upward and insulates it somewhat from the heat.

In the same way, if a galaxy and antigalaxy approach, the first mutual contact at the fringe might produce a flood of energy that will tend to keep them apart and, so to speak, insulate them from each other.

In that case, though, the insulating interaction will be a rich source of gamma ray photons, which would streak through the Universe generally. The gamma rays detected by Explorer XI could conceivably be derived from such sources and could be a signal that the Universe contained as many antigalaxies as galaxies.

And how would that fit the big bang theory?

A contracting Universe containing as many antigalaxies as galaxies ought to undergo more and more mutual annihilation as the galaxies approach, and when the cosmic egg is formed it may consist of gamma rays only. Perhaps it is the radiation pressure of these photons that eventually forces the big bang itself, and in the first ravening moments of expansion, the gamma rays may give rise to an equal quantity of particles and antiparticles.

Again, these particles and antiparticles must be visualized as being separated, or they will eventually fall back into mutual annihilation and gamma rays. We might postulate that for every particle formed on one side of the exploding cosmic egg, a balancing antiparticle is formed on the other side. The Austrian-American physicist Maurice Goldhaber (1911–   ) has indeed suggested something of this sort, visualizing the formation of a Universe of matter and an Antiuniverse of antimatter (which he calls a "cosmon" and an "anticosmon," respectively).

Do the two intermingle in the course of expansion to form

a combined Universe of galaxies and antigalaxies in equal number? If so, we might picture a pulsating Universe, in which there is a merger of matter and antimatter in the course of each contracting phase and a separation in the course of each expanding phase.

Or perhaps the Universe and Anituniverse experience a mutual gravitational repulsion, separating entirely so that our Universe is made up of matter only. Perhaps, in the pulsating model of the Universe, the Universe and Antiuniverse merge in the process of contraction to form a single cosmic egg and separate again in the process of expansion.

Or perhaps (and here is my own personal contribution to the speculation on the problem — one which, to my knowledge, has never been suggested elsewhere) the Universe and Antiuniverse are permanently separated and pulsate in balance. One expands while the other contracts and vice versa.

In such a "double-Universe model," two apparent asymmetries of our own Universe would be wiped out. The double Universe would, *as a whole,* be perfectly balanced in matter and antimatter, while leaving our own Universe to be virtually pure matter. Again, the double Universe would be perfectly balanced in radial motion. The Universe and Antiuniverse taken together would be essentially static, while our own Universe would be left, at this point in the cycle, to its expansion.

CHAPTER **18**

Radio Astronomy

The Sun

While the cosmic rays, X-rays, and gamma rays that reach Earth from outer space are intensely interesting to astronomers, the true breakthrough of the mid-twentieth century came at the other end of the spectrum — in the long-wave, feebly energetic, radio wave portion.

There were two chief reasons for this. First, the atmosphere, which is transparent to visible light but opaque to most other portions of the electromagnetic spectrum, happens to be transparent also to a broad band in the very short-wave radio, or microwave, region. Astronomers were thus offered a second "window" into the heavens. Any microwave radiation from the skies could therefore be studied at leisure from the Earth's surface. There would be no absolute need to send instruments aloft in balloons or rockets.

Secondly, the use of radio waves in wireless communication led to the eventual development of refined techniques for receiving and amplifying weak radiation of this sort.

Indeed, the possibility of radio waves from the heavens occurred to scientists quite early in the game. It was only a few years after the discovery of the radio wave region of the spectrum that speculation arose as to the possibility of detecting radio emission from the Sun. Among others, the English physicist Oliver Joseph Lodge (1851–1940), a pioneer in radio communication, tried to detect solar radio waves about 1890, but failed. Efforts in this direction then languished for a generation, and when success finally came, it was by accident.

The discovery was registered by the American radio engineer Karl Jansky (1905–1950) who, in 1931, was engaged in the purely nonastronomic problem of countering the disruptive effects of static in radio communication. There was one source of static he could not at first pin down, and which he decided, at last, was due to interference from an influx of very short radio waves from outer space. He published his findings in 1932 and 1933, but his papers roused little interest among astronomers.

The only person to take fire, in fact, was another American radio engineer, Grote Reber (1911–   ). In 1937, he built a "radio telescope" in his back yard, a parabolic device thirty-one feet in diameter, designed to gather the weak microwave radiation from space over a sizable area and concentrate it on a receiving device at the focus of the parabola. For several years, Reber painstakingly located the radio sources in the sky. He was the first and, for quite a while, the only "radio astronomer." He published his first paper on the subject in 1940.

Astronomers only slowly began to grow interested. One trouble was that the very short radio waves from the sky were much shorter than those ordinarily used in radio communication, so that radiation from space did not ordinarily interfere with radio reception and, in this fashion, force itself on man's consciousness. Furthermore, technology had not yet developed efficient ways of handling such short radio waves.

In the late 1930's, however, key steps were being taken that eventually broke the dam. Great Britain and the United States were developing "radar" — a device whereby a beam of microwaves was sent outward with the expectation that it would strike an obstacle and be reflected in such a way that a microwave echo might be detected. From the angle from which the echo was received, the direction of the obstacle would be determined. From the time-lapse between emission of the original beam and reception of the echo, the distance of the obstacle could be determined (since microwaves travel at the velocity of light). Radar was ideal for detecting distant objects accurately and quickly, particularly under conditions in which ordinary optical methods were inadequate. It worked as well by night as by day;

and whereas clouds and fog were opaque to light, they were virtually transparent to microwaves.

Radar was used by the British to give early warning of the arrival of German planes and was a crucial aid in winning the Battle of Britain. Any interference with the working of radar was therefore bound to be of the most intense interest to Great Britain and its allies, and in 1942, such interference occurred. The entire radar system was jammed severely by a flood of extraneous microwave radiation. Britain's warning system was temporarily useless. If this was deliberate jamming on the part of the Germans, the consequences could be most serious. Investigation revealed, however, that it was the result of the giant Solar flare, which happened to give the first indication of the existence of Solar cosmic rays (see page 241). The flare had, apparently, sent out a flood of microwaves toward the Earth that had succeeded in handily drowning the man-made radiation that fed the radar system. Thus, it was discovered that the Sun radiated in the radio wave region of the spectrum. By the time World War II was over, astronomers were ready to turn to "radio astronomy" in earnest.

Once the radio wave spectrum of the Sun was placed under study by techniques that had reached high refinement in connection with radar technology, it was quickly apparent that the Sun was emitting far more microwave radiation than could be accounted for by its surface temperature. Some wavelengths were being emitted at intensities that would only be produced at temperatures of 1,000,000° C. or so. These, of course, were emitted by the corona, which was hot enough to radiate X-rays as well (see page 249).

The Sun also emitted bursts of high-intensity microwave radiation in connection with sunspots, flares, and other disturbances.

The Planets

Nor is the Sun (surprisingly enough) the only microwave source in the Solar system. The planets shine only by reflected light, but some of them emit microwaves of their own with sufficient intensity for them to be detected on Earth.

In 1955, for instance, Jupiter was recognized as the source of certain microwave bursts that had been puzzling observers over a period of five years. Some of the general microwave emission of Jupiter was thermal in origin; that is, it originated simply because Jupiter's surface was at a certain temperature and therefore radiated energy over a broad band of the electromagnetic spectrum—a band that included the microwave region. At certain wavelengths, however, the radio emission was much more intense than could be accounted for in purely thermal fashion. (After all, no one expected Jupiter to have a high-temperature corona as the Sun did). The nonthermal radiation was eventually explained by postulating the existence of a Jovian magnetic field far larger and more intense than that of the Earth, as I shall explain in the next section (see page 268).

Equally interesting were the microwaves received from Venus. These were first detected in 1956 and presented astronomers at once with an interesting discrepancy. The measurement of ordinary infrared radiation from Venus had indicated "surface temperatures" of about $-43°$ C. Microwave radiation, however, bespoke temperatures hundreds of degrees higher, well above the boiling point of water, in fact.

But was this really a discrepancy? The infrared radiation from Venus had to originate in the planet's upper atmosphere. If it originated at or near the solid surface of Venus, the infrared radiation would be absorbed by its atmosphere. The atmospheres of both Venus and Earth are, however, transparent to microwaves. Even Venus' opaque and permanent cloud cover, which has kept its solid surface from ever having been seen by the eye of man, is transparent to microwaves. There is a good chance therefore, that the temperature indicated by infrared radiation is at the low level one would naturally expect of the upper atmosphere, while the temperature indicated by microwave radiation is that of the actual solid surface.

Still a surface temperature well above boiling water is rather surprising for Venus. Could it be that the microwave radiation is not thermal in origin but is, at least in part, produced by a magnetic field as in the case of Jupiter? The chances of the latter are small. It was widely believed by astronomers that Venus rotated about its axis very slowly. Since it is also strongly suspected that a magnetic field about a planet originates only when it has a rotation rapid enough to set up eddies within a molten core, it did not seem likely that slowly rotating Venus could have a significant magnetic field.

The matter was settled by the Venus-probe, Mariner II, a well-instrumented rocket which passed within 21,600 miles of Venus in December 1962. No significant magnetic field was detected. If Venus had one at all, it could not be more than a hundredth as intense as that of the Earth. Mariner II data also showed the radiation was not coming from Venus' ionosphere but from its surface. The microwave emission must therefore be thermal in origin, and Venus' surface must be hot. Its intensity as measured from Mariner II indicates a surface temperature for Venus of approximately 400° C.

The Solar system can also yield information through *reflected* microwave radiation.. The first example of this came in 1945 when microwaves were bounced off meteorite showers. Such showers could in this way be detected and studied even in broad daylight when they are ordinarily invisible. To use obstacles still farther away as microwave reflectors required only refinements in technique—the ability to send out very strong impulses and to detect and amplify very weak echoes, from amid environmental radiation of the same nature ("noise").

The Moon, for instance, could be used as an obstacle from which to reflect a microwave beam, and this was accomplished for the first time in 1946. In 1958, echoes were received from Venus, in 1959 from the Sun, and since then from other members of the Solar system such as Mercury, Mars, and perhaps even Jupiter.

The time-lapse between emission and echo-return can be used to determine planetary distances and, indeed, reflections from Venus gave a new and unprecedentedly accurate method of determining the scale of the Solar system. It represented a significant improvement over the parallax determinations of the asteroid, Eros, a generation before (see page 20).

Furthermore, microwave reflections can yield information concerning the nature of the reflecting surface. If a reflecting body were a smooth and perfect sphere, only the portion directly facing the Earth would return an echo. However, if the surface is rough and uneven, sloping ground would be expected to return an echo toward the Earth in regions where no echo would be returned if the ground were smooth. This sloping ground would be a trifle farther from the Earth, however, because of the Moon's curvature, and the echo would be smeared out somewhat, lasting longer than the original emitted pulse. The microwave echoes would also be distorted in certain fashions by Doppler effect if the reflecting object is rotating.

Naturally, much that radar can tell us about the Moon's surface can be checked by evidence of sunlight reflected from that surface. This is not so in the case of Venus, where the solid surface is hidden from us optically, but where that surface can be touched by cloud-piercing microwaves. Thus, in 1965, microwave reflections seemed to indicate the presence of at least two huge mountain ranges on the surface of Venus, one running north and south, the other east and west.

Even more interesting was the question of Venus' rotation. Since nothing could be seen on the globe of that planet but a featureless cloud cover, there was no way of determining accurately the period of its rotation. As late as 1962, far more was known about the rotation of distant Pluto than about the rotation of our nearest planetary neighbor in space. Many guesses, or estimates from inadequate data, had been made, and the most popular suggestion was that the period of rotation was equal to that of the planet's revolution about the Sun—225 days.

If this last situation were so, however, one side of the planet should face the Sun perpetually and the other face away from it just as perpetually (as one side of the Moon faces the Earth perpetually and the other side faces away). One would expect, then, that the "sunside" of Venus would be exceedingly hot, while the "nightside" would be exceedingly cold.

Microwave emissions, rather surprisingly, seemed to indicate, however, that the temperature of Venus' surface did not vary as much as might be expected if there were indeed a sunside and a nightside. One might have to suspect the existence of strong winds that served to carry heat from the former to the latter. Or else,

Microwave reflection

radar pulse lost

the Moon

radar pulse reflected back to Earth

2,160 miles

perhaps Venus' period of rotation was not quite equal to its period of revolution so that there was no sunside or nightside, but instead (as on Earth) every point on the planetary surface was exposed to the periodic presence and, later, absence of the Sun.

The latter suggestion is borne out by microwave reflections. Rather to the surprise of astronomers, Venus was found in 1962, to have a period of revolution of 247 days in the *retrograde* direction. That is, its surface turned in the clockwise direction when viewed from high above its north pole, rather than counterclockwise as in the case of Earth and almost all the other planets. To put it another way, Venus' surface rotated from east to west rather than from west to east as in our case. The period of revolution, combined with the motion of the planet about the Sun means that from any given point of the planet's surface, the Sun would be seen (if the clouds did not exist) to rise in the west and set in the east about twice each planetary year.

Why Venus should rotate in the retrograde direction and why it should be so hot now exercises astronomic thinking but, again, better unanswered questions than no questions.

Nor is Venus the planet about which notions concerning rotation have had to be revised because of the new microwave techniques. In 1965, the planet Mercury was shown by microwave reflection to have a period of rotation that was not equal to its eighty-eight day period of revolution about the Sun. This was even more surprising than the case of Venus, for Mercury has no cloud cover and its surface can be seen (albeit with difficulty in view of its closeness to the Sun), so that its period of rotation can be observed directly by following the shifting pattern of its surface features. As long ago as the 1880's, the Italian astronomer Giovanni Virginio Schiaparelli (1835–1910) had studied those surface features and had maintained that it rotated once per revolution. This had been accepted for eighty years.

When the microwave data came in, Mercury's surface was reobserved with painstaking closeness and Schiaparelli was found to have been wrong. The mistake was, however, understandable. Mercury rotates in 58½ days, or just about two-thirds of its year. The particular face it shows to the Sun at one perihelion, it therefore shows again at the second return to perihelion, the fourth, the sixth, and so on. Anyone making observations separated by some multiple of two revolutions would see the same surface features in the same place and could be pardoned for supposing that the planet rotated once each period of revolution since that is exactly what one would observe in that case, too.

The Stars

Microwave emissions from the bodies of the Solar system, however interesting, are by no means all there is to radio astronomy. Indeed, the very first observations made in the field dealt with microwave sources lying far beyond the Solar system.

Jansky's first observations were made at a period when the Sun was relatively quiet and was emitting microwaves in comparatively minor quantities. It was not therefore the source detected by him. Jansky thought it was, at first, since the source traversed the sky with the Sun. However, as the days passed he noted that the source gained on the Sun by four minutes a day. That meant that the source was maintaining a fixed position with respect to the stars, and as it turned out it was located in the direction of the constellation Sagittarius and is called "Sagittarius A." There was no question that the source was the core of the Galaxy which, judging from the microwave data, is 10 parsecs in diameter and contains 100,000,000 stars.

This was a discovery of the first importance. The clouds of dust that forever blocked off the center of the Galaxy from optical study were quite transparent to microwaves. If we could never see the Galactic nucleus in the ordinary way, we could "see" it by microwaves.

Reber, in the course of his lonely investigations of the microwave emissions received from the heavens, plotted what might be called the "radio sky," the background intensities of microwave emission from place to place. The chief feature of the radio sky was the band of high emission that marched along the Milky Way, with its greatest intensities in the direction of the Galactic nucleus and falling off on either side to a minimum in the direction opposite that of the Galactic nucleus.

However, the plane of the Milky Way did not by any means include all there was to the radio sky. There were knots of high intensity here and there, even in places far removed from the Milky Way. Most of these radio sources could not be identified, at first, with any visible feature, but it was plain they could not represent ordinary stars like our Sun. If stars generally emitted microwaves only as the Sun did, the Galactic center could not radiate enough microwaves to reach us with the intensity they do.

The unusual nature of the radio sources was made perfectly plain by the fact that one of them ("Taurus A") was quickly identified with the Crab Nebula. I have already discussed that object as a source of X-rays and cosmic rays (see page 251). It proved also to be a source of microwaves, the third strongest source outside the Solar system.

At first, one might suppose that the microwave emission of the Crab Nebula, far far stronger in intensity that that from our Sun, was merely the result of the same high temperatures that produced X-rays and cosmic rays. Apparently not, however. If the temperature was high enough to produce microwaves in the intensities observed, the Crab Nebula would have to be much brighter optically. Secondly, while the intensity of microwaves produced by high temperature should decrease as wavelength increases, this does not happen in the case of the Crab Nebula.

A suggestion arose out of findings in the field of nuclear physics. It takes energy to make a moving body change its speed or direction of motion, and the moving body, in response to this, can radiate away some of the energy. This was noted particularly where instruments called "synchrotrons" made use of intense magnetic fields to whirl electrons in circles with energy being poured into them constantly. The electrons emitted "synchrotron radiation" with wavelengths and intensities that depended on their energy content.

In 1953, the Soviet astronomer Iosif Samuilovich Shklovsky (1916–   ) suggested that the Crab Nebula might have an intense magnetic field and that the magnetic lines of force might cause electrons to spiral, emitting synchrotron radiation as they did so, and that this radiation might include microwaves as well as light.

If this were so, then the wave forms emitted would be guided by magnetic lines of force oriented in a fixed manner in space. The wave forms would themselves, therefore, have a fixed plane, and the light from the Crab Nebula would have the property of being "polarized." Astronomers can test whether light is polarized or not, and the Soviet astronomer V. A. Dombrovsky was the first to show that indeed it was, a point quickly confirmed by other astronomers. Views of the Crab Nebula taken through Polaroid filters oriented in given directions, show straight-line features in each case, in directions perpendicular to that orientation. That would be expected in the case of polarized light.

It was the success of this view in connection with the Crab

Nebula that led to the suggestion that some of Jupiter's non-thermal emissions might be synchrotron radiation (see page 269). This meant that Jupiter would have to have a magnetic field some tens of times stronger than the Earth's, and in view of Jupiter's rapid rotation (ten hours as compared with the Earth's twenty-four, although Jupiter has eleven times the diameter of the Earth) that seemed quite likely. By 1960, Jupiter's microwave emissions were found to be polarized in the direction that would be expected if Jupiter's magnetic poles were near its geographic poles, as is true of the Earth.

By that time, the efficiency with which particles were trapped by magnetic lines of force had come to be understood, and the magnetosphere about the Earth (see page 240) had been discovered. A much more intense magnetosphere about Jupiter had to be inferred. Moreover, in 1962, when a nuclear bomb was exploded high above the atmosphere, the Earth's magnetic lines of force were flooded with charged particles which were trapped and which, as they spiraled back and forth about the lines, actually emitted detectable synchrotron radiation. That filled out the picture, and the synchrotron radiation theory of microwave emission by the Crab Nebula (and many other radio sources) came to be generally accepted.

The Crab Nebula is, of course, almost certainly the remnant of a supernova, and other radio sources can also be identified with supernovae known to have exploded in our Galaxy. Those of Tycho and of Kepler are examples. The most powerful radio source of all is not, however, connected with a *known* supernova. It is called "Cassiopeia A," for it occurs in that constellation. Nothing startling corresponds optically to Cassiopeia A; all that shows up are clouds and wisps of gas about 10,000 light-years from us. This gas, on closer study, proved to be ferociously hot and in violent movement. It may well be the remnant of a supernova that exploded as recently as 1700 but went unnoticed because its distance kept it from being particularly bright, at a time when interest in "new stars" was so small that only particular brightness would have brought it to the attention of observers. Another interesting radio source is IC443, also a nebula that may be a supernova remnant, one that is perhaps 50,000 years old. The white dwarfs associated with such nebulosities cannot be seen—again because of distance.

It is reasonable to suppose that the microwave emission of the Milky Way generally might be produced by the supernovae that have appeared within it, but they need not be the only sources. There is a type of red dwarf star that flares up occasionally in an irregular fashion. Presumably these stars brighten by means of flares, like those of the Sun, only more intense, and these flares liberate microwaves as do those of the Sun.· Cooperative work between the English astronomer Alfred Charles Bernard Lovell (1913–   ) and the American astronomer Fred Lawrence Whipple (1906–   ) showed that the intensity of microwave emission did indeed parallel the brightening effect, and these "flare stars" were the first reasonably ordinary individual stars to be identified as radio sources.

The Galaxy

Not all microwave emission from the Galaxy originates in stars, or in the remnants of supernovae. There is also the matter between the stars, the thin interstellar gas that consists mostly of hydrogen. If this hydrogen happens to be heated by some nearby star, the atoms can be pumped full of energy and ionized. This energy can be radiated so that the astronomer can observe luminous clouds and detect spectral lines associated with hydrogen.

This is better than nothing, but not much better, because only a small proportion of the interstellar hydrogen of the Galaxy is heated sufficiently to produce these lines. At least 95 percent of the hydrogen of interstellar space is relatively cold and, radiationally speaking, quiet. (These are "H-I regions" as opposed to the hot and ionized "H-II regions.") Furthermore, the lines emitted by hot hydrogen can only be seen where dark nebulae do not interpose their obscuring clouds, which means that only our own section of the Galaxy can be studied in this manner.

In 1944, however, a Dutch astronomer, Hendrik Christoffel van de Hulst (1918–   ), amused himself with pen-and-paper calculations concerning the behavior of cold hydrogen. (He was forced into this by the fact that the German occupation of Holland during World War II made ordinary astronomic work impossible.) He worked out the manner in which the magnetic fields, associated with the proton and the electron in the hydrogen atom, were oriented to each other. They could both line up in the same direction or in opposite directions. There is a slight energy difference between the two, and every once in a while a hydrogen atom in the less energetic form might absorb a photon of just the right size that happens its way and move into the more energetic form. Again, every once in a while a hydrogen atom in the more energetic form might emit a photon and sink to the less energetic form. The energy taken up or given off is so small that only very unenergetic photons are absorbed or produced, photons in the microwave region with wavelengths of about 21 centimeters.

The emission or absorption of a 21-centimeter microwave photon by any one hydrogen atom ought to take place only very rarely—once in 11 million years, on the average—but there are so many hydrogen atoms in space generally that a steady and perhaps detectable drizzle of such events must be taking place.

After the war was over, astronomers began to search for some evidence of this, and in 1951, the Swiss-American physicist Felix Bloch (1905–   ) and the American physicist Edward Mills Purcell (1912–   ), working independently, detected the 21-centimeter radiation. Hydrogen absorption at the 21-centimeter mark was also detected eventually.

Now a method existed for detecting the interstellar hydrogen in space and for telling where it was present in relatively thick profusion and where it was not. Furthermore, since the photons were those of microwaves and not of visible light, they could penetrate dust clouds easily and astronomers could "see" the interstellar hydrogen in many parts of the Galaxy which were optically invisible.

Since it was to be expected that the dust and gas of the Galaxy are concentrated in the spiral arms, the mapping of the 21-centimeter sources ought to give one an idea of the spiral structure of our Galaxy.

Actually, before the detection of the 21-centimeter radiation, attempts had already been made to trace the spiral arms by means of the luminous clouds of hot hydrogen. They would surround the particularly hot Population I stars that formed in the dusty arms. The line of blue-white giants, lighting up the hydrogen all about for light-years, would trace out the arms. Using this technique, the American astronomer William Wilson Morgan (1906–   ) and his associates prepared a map of the spiral arms of our Galaxy in 1951.

Sections of three separate arms were marked out. One of these included features in Orion and was therefore called the "Orion Arm." It includes our own Sun. Closer to the Galactic center than that is the "Sagittarius Arm" and farther out from the center than the Orion Arm is the "Perseus Arm."

Further investigations of this sort made the map more intricate, but then the technique became obsolete in the light of the 21-centimeter radiation. Suddenly, it became possible to work for much larger distances within the Galaxy and in much greater detail. Maps were prepared of the spiral structure of the Galaxy, and one could begin to think of it, schematically, as a rather symmetrical double spiral.

Nor is the cold, neutral hydrogen gas of the Galaxy static. Studies by Oort and Van de Hulst seem to indicate that the hydrogen flows outward from the center to the outskirts of the system at a surprisingly rapid rate. Oort estimates that the quantity of hydrogen transported each year from the center outward is equal to the mass of the Sun. This flow of gas outward along the spiral arms may, according to some speculations, serve to keep the arms in being, maintain their rich supply of gas and their ability to form new stars. On the other hand, it is difficult to see how the source of hydrogen at the center persists. It should have run dry long ago, unless there is a general circulation by which the supply at the center can be replenished, perhaps at the expense of a gigantic "halo" of hydrogen gas that seems to encompass the Galaxy generally. What keeps the hydrogen circulation in being is not known as yet.

Spiral arms of the Galaxy (schematic)

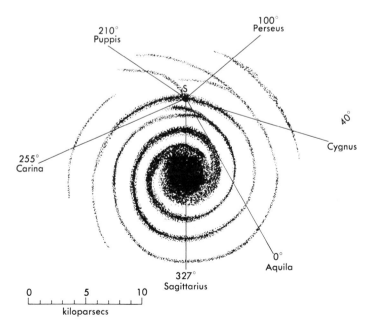

Other galaxies also have their supply of hydrogen, of course, and the amount may depend on the type of galaxy. Spiral galaxies have more hydrogen, it seems, than elliptical galaxies do. Study of the 21-centimeter radiation makes it appear that more open spiral galaxies have more interstellar hydrogen than tighter ones have, while irregular galaxies have most of all. What bearing this has, if any, on the question of the evolution of galaxies, is, as yet, uncertain.

Naturally astronomers are anxious to find other types of microwave radiation that might be useful in studying interstellar gas, and for that reason considered deuterium (hydrogen-2). Hydrogen is by far the most dominant component of interstellar gas, and a certain small percentage of its atoms must be hydrogen-2. The hydrogen-2 atom differs from the ordinary hydrogen-1 atom in having a neutron added to the proton of the nucleus. The magnetic field of the proton-neutron nucleus ought to interact with the magnetic field of the electron as in the case of the simple proton nucleus, theory predicted, and a microwave with a 91-centimeter wavelength should be emitted. Strong radio sources such as Cassiopeia A were combed for the telltale wavelength, and in 1966 it was detected by astronomers at the University of Chicago. The amount of deuterium in the universe seems, now, to be about 5 percent that of ordinary hydrogen.

Next to hydrogen, the most common element in the Universe is helium. Microwaves from helium at wavelengths calculated from theoretical considerations by the Soviet astronomer, N. S. Kardashev, were also detected in 1966.

Following hydrogen and helium, the most common component of interstellar gas is oxygen. An oxygen atom can combine with a hydrogen atom to form a "hydroxyl group." This combination would not be stable on Earth, for the hydroxyl group is very active and would combine with almost any other atom or molecule it encountered. It would, notably, combine with a second hydrogen atom to form a molecule of water. In interstellar space, however, as Shklovsky pointed out in 1953, where the atoms are spread so thinly that collisions are few and far between, a hydroxyl group, once formed, would persist undisturbed for long periods of time.

Such hydroxyl groups would, calculations showed, emit or absorb four different wavelengths of microwaves. Cassiopeia A was searched for these and in 1963, two of them—absorption lines in the neighborhood of the 55-centimeter mark—were detected by a team of radio engineers at Lincoln Laboratory of M.I.T.

The hydroxyl absorptions turned out to be unexpectedly useful. Since the hydroxyl group is some seventeen times as massive as the hydrogen atom alone, it is more sluggish and moves at only one-fourth the velocity of the hydrogen atom at any given temperature. In general, movement blurs the line, and the hydroxyl absorption line is much sharper than is that of hydrogen. It is easier to tell whether the hydroxyl line has shifted slightly toward shorter or longer wavelength and to determine, in that way, the radial velocity of the gas cloud.

Furthermore, the ratio of hydroxyl groups to hydrogen atoms varies, for some reason, from place to place, and the relative quantity of hydroxyl groups seems to increase rapidly as one investigates objects closer and closer to the Galactic center and nearer the hot H-II regions. Astronomers hope, therefore, that the hydroxyl group map of the Galaxy will pinpoint the center sharply and make possible a clearer interpretation of the events taking place there.

In 1967, it was discovered that some hydroxyl-rich regions were no larger in size than our Solar system. The suggestion was at once made that they represented stars in the process of formation; stars that were still cool enough to allow hydrogen and oxygen atoms to hang together as hydroxyl groups. The coolest, youngest stars were thought to be the infra-red giants (see page 133). They were promptly investigated and several were indeed found to be associated with hydroxyl radiation, which increases the plausibility of the suggestion.

Pulsars

In the 1960's, astronomers grew interested in the manner in which microwaves from a particular source managed to change in intensity. The change was not surprising if it were slow, but about 1964 it was noted that some sources changed the intensity of their microwave emission with astonishing rapidity. It was almost as though there were a microwave-twinkle.

This caused astronomers to design radio telescopes capable of catching very short bursts of microwave-energy. They felt this would make it possible to study these fast changes in greater detail. One astronomer making use of such a radio telescope was Anthony Hewish at Cambridge University Observatory.

He had hardly begun operating the telescope when he detected bursts of microwave energy from a place midway between Vega and Altair. He was astonished for though he had hoped for quick bursts he had expected nothing this brief; each burst lasted only 1/30 of a second. What's more it was even more astonishing that they followed one another with remarkable regularity at intervals of 1 1/3 seconds.

Saying 1 1/3 seconds is hardly fair. The intervals were so equally and accurately spaced that the value could be worked out to a hundred-millionth of a second. The period of this first example of a new phenomenon was 1.33730109 seconds.

It might seem odd that this radio source should be picked up so quickly by the new instrument. How was it that ordinary radio telescopes had missed it?

The individual bursts, to be sure, are so energetic that they are momentarily very bright and easy to detect—but only momentarily. A radio telescope not designed to detect very short bursts, only picks up the average emission from the spot and that is 3 percent or less of the maximum brightness, and that went unnoticed.

Hewish hadn't the faintest idea what this odd microwave phenomenon might represent. Since the microwaves seemed to emerge from a point in the heavens, he considered it as representing some kind of star. Since the microwaves emerged in short pulses, he thought of it as a kind of "pulsating star." This was shortened almost at once to "pulsar" and it is by that name that the new object came to be known.

One should say objects and use the plural for once Hewish found the first, he searched for others, and by February 1968, when he announced the discovery, he had located four. Other astronomers began to search avidly and more were quickly discovered. In two more years, nearly forty more pulsars were located.

Two-thirds of them are located very close to the Galactic equator, which is a good sign that pulsars generally are part of our own Galaxy. (There is no reason to suppose they don't exist in other galaxies, too, but at the distance of other galaxies they are probably too faint to detect.) Some may be as close as 100 light-years or so.

All the pulsars are characterized by extreme regularity of pulsation, but of course the exact period varies from pulsar to pulsar. One had a period as long as 3.7 seconds. Then, in November 1968, astronomers at Green Bank, West Virginia, detected a pulsar in the Crab Nebula, which had a period of only 0.033089 seconds. It was pulsing 30 times a second.

Naturally, the question is, what can produce such short flashes in such fantastic regularity. So stunned were Hewish and his fellow-astronomers at the first pulsars, that they wondered if it might not be signals from some intelligent life-forms far out in space. It quickly appeared, though, that the wavelengths present in the pulses were easily obscured by microwave radiation in the Galaxy generally. Surely, no intelligent life-form capable of producing such pulses would be so unintelligent as to select such wavelengths. Besides an improbable quantity of energy is involved. To produce the pulses would require 10 billion times the total quantity of power man can produce. It doesn't seem that much power would be wasted just to send out very regular signals, and the more pulsars are detected the less likely it seems that so many different life-forms are all zeroing their signals in on us. The theory was quickly dropped therefore.

But something must be producing them; some astronomical body must be undergoing some change at intervals rapid enough to produce the pulses.

You might argue, for instance, that a planet was circling a star in such a way that once each revolution it moved behind the star (as seen from the direction of Earth). Each time it emerged from behind the star it emitted a powerful flash of microwaves for that reason. Or else the planet rotated and each time it did so some particular spot on its surface which leaked microwaves in vast quantity would sweep past our direction and we would receive a pulse.

To do this, however, a planet must revolve about a star or rotate about its axis in a period of anywhere from 4 seconds down to 1/30 seconds and this is unthinkable.

For pulses as rapid as those of pulsars, some object must be rotating or revolving at enormous velocities and that requires enormous gravitational fields. This instantly brought white dwarfs (see page 155) to mind.

Theoreticians got busy at once, but try as they might, they could not find any way of visualizing one white dwarf circling another, or rotating about its axis, or pulsating, with a period quickly enough to account for pulsars. White dwarfs were too large and their gravitational fields were too weak.

Something smaller and denser than a white dwarf was required and Thomas Gold suggested that a neutron star (see p 252) was involved. Once before astronomers had speculated about neutron stars, in connection with X-ray sources, but that brief flare-up had died down. Now the notion was in the air again.

Gold pointed out that a neutron star was small enough and dense enough to be able to rotate about its axis in 4 seconds or less. What's more, it had already been theorized that a neutron star would have a magnetic field and magnetic poles that would be near the geographic equator. Electrons would be held so tightly by the neutron star's gravity that they could emerge only at the magnetic poles. As they were thrown off, they would lose energy in the form of microwaves. This means there would be a steady sheaf of microwaves emerging from two opposite points on the neutron star surface.

If, as the neutron star rotates, one of those points happens to move about the axis of rotation in such a way that the sheaf of microwaves sweeps past our direction, then we will detect a short burst of microwave energy once each revolution. If this is so we would detect only pulsars which happen to rotate in such a way as to sweep one of the magnetic poles in our direction. Some astronomers estimate that only one neutron star out of a hundred would do so and guess that there might be as many as 100,000 neutron stars in the Galaxy, but that only 1000 would be detectable from Earth.

Gold went on to point out that, if his theory were correct, the neutron star would be leaking energy at the magnetic poles and its rate of rotation would be slowing down. This meant that the faster the period of a pulsar, the younger it was likely to be and the more rapidly it might be losing energy and slowing down.

The most rapid pulsar known is in the Crab Nebula and it might well be the youngest, since the supernova explosion that would have left the neutron star behind took place only a thousand years ago.

The period of the Crab Nebula pulsar was studied carefully, and it was indeed found to be slowing up, just as Gold had predicted. The period was increasing by 36.48 billionths of a second each day, and at that rate it will have doubled in 1,200 years. The same phenomenon has been discovered in other pulsars whose periods are slower than that of the Crab Nebula pulsar and whose rate of slowing is also slower. The first pulsar discovered, now called CP1919, has a period 40 times as long as that of the Crab Nebula pulsar, and it is slowing at a rate that will double its period only after 16 million years. As a pulsar slows, its pulses become less energetic. By the time the period has passed four seconds in length, the pulsar becomes too weak to be detectable. Pulsars probably endure as detectable objects for tens of millions of years, however.

As a result of these studies of the slowing of the pulses, astronomers are now pretty well satisfied that the pulsars are neutron stars.

Sometimes a pulsar will suddenly speed up its period very slightly, then resume the slowing trend. This was first detected in February 1969, when the period of the pulsar Vela X-1 (found amid the debris of a supernova that blazed up 15,000 years ago) was found to alter suddenly. The sudden shift was called, slangily, a "glitch," from the Yiddish word meaning "to slip."

Some astronomers suspect glitches may be the result of a "starquake," a shifting of mass distribution within the neutron star that will result in its shrinking its diameter by a centimeter or less. Or perhaps it might be the result of some sizable meteor plunging into the neutron star and adding its own momentum to that of the star.

There was, of course, no reason why the electrons emerging from the neutron star should lose energy only as microwaves. It should produce waves all along the spectrum.

Neutron stars should, for instance, emit X-rays, too, and the Crab Nebula neutron star does, indeed, emit them. About 10 to 15 percent of all the X-rays the Crab Nebula produces are from its neutron star; it was the other 85 percent or more that come from the turbulent gases that obscured this fact and disheartened those astronomers who hunted for a neutron star there in 1964.

A neutron star should produce flashes of visible light, too. In January 1969, it was noted that the light of a dim sixteenth-magnitude star within the Crab Nebula does flash on and off in precise time with the microwave pulses. The flashes and the period between them are so short that special equipment was required to catch them. Under ordinary observation the star seems to have a steady light.

The Crab Nebula neutron star was the first optical pulsar discovered, the first visible neutron star. In 1977, the Vela pulsar was detected optically. It was the second.

One important property of the neutron star is its mass. In 1975, the mass of a neutron star was determined for the first time when Vela X-1 turned out to have a mass 1.5 times that of the Sun. This was slightly over Chandrasekhar's limit and no white dwarf could be that massive (although neutron stars with masses under

Chandrasekhar's limit are possible).

The mass of Vela X-1 was capable of being determined because that neutron star is part of a binary. Its companion is a massive star of the main sequence, one with 30 times the mass of our Sun. Eventually, in a million years or less, the companion of Vela X-1 will go supernova in its own right, and there may then be two neutron stars rotating about a common center of gravity.

A neutron star has an extremely intense gravitational field near itself, of course. A neutron star with the mass of the Sun and a diameter of 9 miles (14 kilometers) would have a gravitational pull on its surface of 280,000,000,000 times that we feel on the Earth's surface.

To get away from a neutron star that is capable of pulling on an object that intensely, it requires enormous speeds. The escape velocity from a neutron star is something like 125,000 miles per second (200,000 kilometers per second). This is about two-thirds of the speed of light, so that we might imagine that even light and its companion radiations such as microwaves and X-rays would have some difficulty getting away. The Einstein red-shift would rob it of much of its energy.

Black Holes

Even the neutron star does not represent the ultimate.

What if a really massive star were to explode and the condensing remainder was several times the mass of our Sun. It might be that even neutrons couldn't withstand the increasing intensity of the gravitational field. The neutrons might smash, also. What then?

In 1939, J. Robert Oppenheimer (1904–1967) was working out the theoretical implications of neutron stars, and he took up the matter of increasing mass. It turned out that a neutron star could not have a mass of more than 3.2 times that of the Sun. If an object more massive than that collapsed, then even tightly packed neutrons would give way and the star would continue to collapse *past* the neutron-star stage.

Under those circumstances, there would remain nothing—nothing at all—to stem the collapse. The matter of the collapsing star would contract to zero and the density would rise to unlimited heights.

The result would be an object with a surface gravity so enormous that the escape velocity would rise beyond the velocity of light. As you move away from the zero-volume center of contraction (the "singularity") the escape velocity drops and, at a certain distance, it drops to merely the velocity of light.

If we imagine the singularity surrounded by a sphere at every point of which the escape velocity is equal to the velocity of light, then we might consider that the size of the collapsed star. The distance from the center to the sphere is the "Schwarzschild radius" because it was first calculated by the German astronomer Karl Schwarzchild (1873–1916).

If a star as massive as our Sun were to collapse to nothing, the Schwarzschild radius would be just under 2 miles, or about 3 kilometers.

Since nothing can surpass the speed of light, nothing can emerge from within the Schwarzschild radius. Objects can fall in but they can't come out. Such a collapsed star is like a bottomless hole in space. Since even light can't emerge, it is black. The American physicist John Archibald Wheeler (1911–   ) called the object a "black hole" and that name caught on.

Even though black holes exist in theory, do they actually exist in practice? If no light comes out of them, how can they be detected? We might detect their gravitational pull, if they were close enough, but they would have to be very close.

Fortunately, there are times when radiation *is* given off by a black hole; not by the black hole itself but by matter falling into it.

If a black hole is party of a binary system, and if the other star is a large main-sequence star, then it is possible for matter from the ordinary star to drift toward the black hole and circle it. The circling matter is an "accretion disk." As the accretion disk circles and gradually falls into the hole, it gives off energetic radiation such as X-rays.

In 1969, an X-ray detecting satellite was launched from the coast of Kenya on the fifth anniversary of Kenyan independence. It was named Uhuru from the Swahili word for independence. It multiplied knowledge of X-ray sources, detecting 161 such sources.

In 1971, Uhuru detected a marked change in X-ray intensity in Cygnus X-1. When this source was looked at closely, it was found to be very near a large, hot blue star, HD-226868, some 30 times as massive as our Sun. HD-226868 was part of a binary system. It was circling about some sort of center of gravity with a period of 5.6 days. From the position of the center of gravity, it seemed that the other part of the binary system was a star that was 5 to 8 times as massive as the Sun and was a source of intense X-rays. The other star could not be seen, however, and for a star that massive not to be seen gives astronomers a strong feeling that it is a black hole.

It also seems that the centers of many galaxies are the sites of strong radiation of various sorts, even the center of our *own* Galaxy. At the centers of galaxies, the stars are very densely packed and it may be that once a black hole is formed there, it is easy for it to pick up new matter and grow.

Eventually, these galactic black holes may be as massive as millions, or even billions, of ordinary stars and may be gulping down whole stars at a swallow. Eventually, perhaps, black holes will have hollowed out the central regions of all galaxies, and even of globular clusters. The outskirts of galaxies, however, should remain safe.

If black holes were formed only by the collapse of large stars, then all black holes should be more massive than the Sun. Some astronomers think that in addition to the really huge black hole at the center of the Galaxy, there may be as many as a billion star-sized black holes scattered here and there throughout the Galaxy.

Could there be black holes that had masses less than that of stars? Possibly. According to theory, there could be black holes of any size. Even a single proton, if it was squeezed down to a much smaller size than it ordinarily had, would become a very tiny black hole.

But how could small objects be squeezed so hard as to form black holes? Large stars can collapse to black holes because they have enormous gravitational fields that do the compressing. What can compress small objects?

In 1971, the English astronomer Stephen Hawking suggested that at the time of the big bang the expanding gases may have compressed parts of themselves sufficiently to form black holes of all sorts of sizes, including some with masses no larger than that of Earth, or even with masses far smaller. These Hawking called "mini-black holes."

There is no evidence yet that such mini-black holes exist, but Hawking was able to show that black holes *do* lose matter, but only very, very slowly. An ordinary black hole such as Cygnus X-1 would lose mass so slowly that even if it were to gain no more mass from the outside universe, it would take countless trillions of years before its mass was reduced by, say, 10 percent.

However, the less massive a black hole, the more rapidly it can lose mass—and as it loses mass it has less mass and loses further mass even more rapidly until, finally, it becomes so small that all the little mass it has left goes at once and it explodes. In the process of explosion, it would release gamma rays of a certain kind.

If mini-black holes existed, some would be so tiny that they would have exploded billions of years ago. Others would be so large that they wouldn't get to the explosion point for billions of years. Some, however, that were formed at the big bang 15 billion years ago might have been formed with just the size that would enable them to last 15 billion years before exploding. In that case, they would be exploding right now. Astronomers are watching for just those telltale gamma-ray signals that would signal the death of a mini-black hole. So far they have found none.

19

The Edge of the Universe

Colliding Galaxies

By the mid-1950's nearly 2000 separate radio sources had been marked out in the sky. A number of them were obviously part of the Milky Way complex of general microwave emission. About 1900 were not, however. These were not smeared out over a sizeable area as the microwave emission from the Milky Way was. Rather, as the Australian astronomer John G. Bolton first showed in 1948, they were point sources, with microwaves emerging from small areas in the heavens.

It seemed logical to suppose such microwaves arose from stars which, for one reason or another, radiated heavily in that region of the spectrum. Indeed, Bolton called them "radio stars."

It was clear, of course, that such radio stars, if they really existed, could not be ordinary stars, but were probably remnants of supernovae. Certainly this was true of the Crab Nebula, which was the third brightest radio star, and somewhat less certainly, of Cassiopeia A, which was the brightest.

(There is some speculation to the effect that Cassiopeia A may be the remnant of a supernova that exploded only 300 years ago, but was not visible when it did so because of dust clouds between it and ourselves. Microwave radiation could penetrate those dust clouds easily. —Another radio source in Cassiopeia (Cassiopeia B) may be the remnant of Tycho's Supernova, (see page 165).

The trouble was, though, that very few radio stars could be associated, in those early years of radio astronomy, with any object that was visible optically, not even with faint nebulous patches as in the case of Cassiopeia A. Part of the trouble was that the sharpness with which any object is viewed is dependent on the wavelength of the radiation by means of which it is viewed. Radiation with long wavelength makes for fuzzier vision. Microwaves are roughly 400,000 times longer in wavelength than are the light waves by which we ordinarily see the stars, and "vision" is correspondingly fuzzier. The result is that viewing the sky by microwaves is like viewing it through optical equipment that is badly out of focus. Instead of a sharp point, we get a dim patch of fog, and where in that fog is the actual point we are trying to see?

Radio astronomers had to locate their radio source as well as they could, pin it down to a certain small area—small to the naked eye, but gigantic to a powerful optical telescope—and then try to see if somewhere in that area something suspicious is visible in the light-wave region. If there is, every attempt is made to sharpen the microwave focus and see if it seems to be zeroing in on the suspicious site. Efficiency in this direction increased with passing time as radiotelescopes grew larger and detecting equipment more refined.

But as the years passed, astronomers grew increasingly uneasy over the notion of radio stars. No matter how the position of these objects was boxed in with increasing accuracy, nothing visible within our Galaxy could be pinned down, except in a very small minority of cases. Worse still, the greater the number of radio stars located, the more it became evident that they were spread out all over the sky quite evenly, whereas all objects within our Galaxy from ordinary stars to supernovae remnants were heavily concentrated in the plane of the Milky Way. In fact, the only optically visible objects that were spread out evenly all over the sky were the galaxies. Could it be then that the so-called radio stars were galaxies? Ought one to speak of "radio galaxies" rather than of radio stars?

The first real breakthrough in this direction came in connection with the second brightest radio source, one called "Cygnus A." Microwave emission from its general direction had been noted by Reber in 1944, but by 1948, Bolton had shown it to be one of the radio stars. It was the first source, in fact, that he could identify as sufficiently sharp to warrant being called a radio star. By 1951, the position of Cygnus A had been boxed into an area about 1 minute of arc squared. The problem, then, was to locate something within that square.

Baade studied that square with the 200-inch telescope and spotted an oddly shaped galaxy within it. On closer investigation, it seemed to be not one distorted galaxy, but two galaxies with their nuclei in near contact.

The explanation seemed clear. Two galaxies were in collision! Just as there were catastrophes on the stellar level which, as in the case of the Crab Nebula, resulted in an out-spewing of microwaves, so there were still more colossal catastrophes on the galactic level, with still larger microwave emissions.

And it seemed clear that the colliding galaxies were indeed undergoing colossal travail. When the optical spectrum was finally obtained (a difficult task in view of its faintness), that spectrum showed the lines of highly ionized atoms, lines that could be present only if the temperature were extraordinarily high. (Baade had bet Minkowski a bottle of whiskey that this would be the case, and won.)

Suspicion arose at once that all or almost all the so-called radio stars were actually galaxies in collision, and the search was on for other cases of the sort; or, indeed, for any "peculiar galaxy" —one with some oddity of shape or structure that might indicate an unusual event on a huge scale.

They were found in considerable numbers. More than a hundred "radio galaxies" have now been identified, and many of them are peculiar indeed. There is, for example, galaxy NGC 5128 which seems to be a spheroidal galaxy with a thick band of dust running down its middle. It was suggested that this, too, might represent a galactic collision and that what we saw was a spiral galaxy, viewed edgewise with the dust of its arms obscuring its center, knifing its way through the spheroidal galaxy.

Astronomers calculated the probabilities of galactic collisions and decided they might be much more likely than stellar collisions. Our Sun, for instance, is 860,000 miles in diameter and is 25,000,000,000,000 miles from its nearest neighbor. If this is typical, then the distance between stars is just about 30,000,000 times their diameters. A star moving randomly is much more likely, therefore, to pass through the vast empty spaces between stars than to zero in on the comparatively minute body of a star itself.

Our Galaxy, on the other hand, is 100,000 light-years in

diameter and our nearest large neighbor, the Andromeda galaxy, is 2,300,000 light-years away. If this is typical, then the distance between galaxies is about twenty times their diameter. Space is much more crowded (relatively) with galaxies, than galaxies are with stars, and intergalactic collisions are correspondingly more likely than interstellar collisions.

In the hey day that followed the discovery of the apparently colliding galaxies in Cygnus, it was calculated that in our own neighborhood of the Universe there ought to be five collisions per billion galaxies, while within galactic clusters collisions ought to be even more common. In a 500-galaxy cluster in the constellation Coma Berenices it was estimated that at least two collisions ought to be proceeding at any given moment and that every galaxy in it ought to undergo several collisions in its lifetime.

To be sure, when galaxies collided and passed through one another, there is no question of wholesale stellar collisions. The stars are widely separated compared to their size and one galaxy can pass completely through another without much danger that any of the stars of one would actually collide with any of the stars of the other. Nevertheless, the dust clouds of one are apt to collide and pass through those of the other, and this might supply the actual source of the microwave emission.

That this might be so was evidenced by the fact that as the actual radio source in Cygnus was pinpointed further, it seemed not to arise from the dust-free galactic nuclei in collision, but from two points well on either side of those nuclei—in the spiral arms, presumably, where the dust would be concentrated.

Exploding Galaxies

The life of the colliding theory was merry, but short. For one thing, the question of energy arose to plague astronomers.

An ordinary galaxy, like our own, will emit about 10,000,000,000,000,000,000,000,000,000 (ten thousand trillion trillion) kilowatts of energy in the form of microwaves. This is about the equivalent of a thousand individual radio sources such as Cassiopeia A.

This is a comfortable fact. It is perfectly reasonable to explain the microwave emissions of an ordinary galaxy by supposing it to contain several thousand supernovae remnants. Such a figure is certainly not excessive. The microwave emission of an ordinary galaxy represents only about a millionth of the energy emitted in the form of light and this, too, is easy to accept.

Among the radio galaxies, however, even the weakest pour out into space a hundred times the microwave energies that ordinary galaxies do. Cygnus A radiates a million times as much microwave energy as ordinary galaxies do. Indeed, Cygnus A radiates five times as much energy in the form of microwave as in the form of light. What's more, it produces fantastic quantities of cosmic rays. The cosmic ray energy present in Cygnus A is thought to be equal to seven times that which the Sun could produce if all its mass were turned to energy.

This began to seem more and more puzzling, and to account for such large batches of microwaves alone became more difficult, the more the matter was considered. It turned out, for instance, that the energy of microwave emission of Cygnus A was about equal to the total energies of motion of the supposedly colliding galaxies. It seemed completely incredible that the energy of collision could be converted completely into microwaves. All the mass of the galaxies would have to be brought to a standstill with respect to each other and how could that be done? By tens of billions of stellar collisions? Impossible! And even if that happened, how could all the energy come off as microwaves? Surely much of it would be emitted as radiation in other regions of the spectrum.

Radio sources within Galaxies

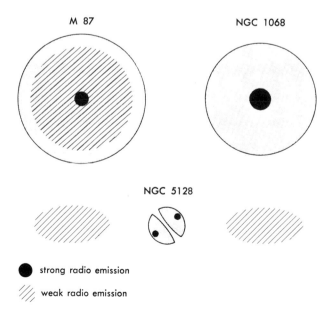

M 87

NGC 1068

NGC 5128

● strong radio emission

⁄⁄⁄ weak radio emission

Furthermore, as the 1950's progressed toward their close, it became more and more generally accepted that the microwave emission from the various radio sources arose from the synchrotron radiation of high-energy electrons trapped in a magnetic field (see page 269). This meant that the kinetic energy of collision would not be converted into microwaves directly but into very high-energy electrons which would then be trapped in a magnetic field. No reasonable mechanism could be advanced for such a conversion of kinetic energy into high-energy electrons.

There was observational evidence against the colliding-galaxy theory, too. As more and more microwave sources were pinned down to galaxies, it became more and more difficult to interpret everything one saw in those galaxies as representing collisions. Indeed, some "peculiar galaxies" might be peculiar in the intensity of their microwave emission, but they were not peculiar at all in shape or appearance. They seemed perfectly ordinary galaxies, abiding in single blessedness with no sign of any collision, and yet they were strong microwave emitters.

An alternate view began to make itself felt. Perhaps it was not a matter of the collision of two galaxies at all, but the explosion of a single galaxy.

For instance, the galaxy NGC 1068 is a weak radio galaxy with microwave emission only about a hundred times that of a normal galaxy, but that emission seems to arise entirely from a small area right in the center of the galaxy. A collision of galaxies, involving dust clouds, would be expected to spread the source over a wider area and certainly not in the dust-free center. An explosion, on the other hand, would be expected to go off right there in the center, where stars are most crowded together and where some sort of catastrophe involving large numbers of stars over a relatively short span of time might most easily take place. If so, we may be viewing the beginning of such a central catastrophe in NGC 1068. The microwave emission is still highly concentrated in that beginning-to-explode center, and it is still weak.

A later stage in the process is perhaps displayed by galaxy NGC 4486, which is better known as M87 from its position on Messier's list. (This happens to be the most massive known Galaxy and may contain as many as 3000 billion stars.) It, too, has an intense microwave source at the center, but it emits microwaves more weakly throughout a halo about that center, one that takes up almost the whole volume of what can be seen optically. It is as though the ravening fury of the central explosion may have spread out for tens of thousands of light-years in every direction, and M87 emits microwaves with a hundred times the intensity of NGC 1068. What is most interesting is that close optical study of M87 shows that a luminous jet is emerging from its center. Can that be material hurtled out into intergalactic space by the fury of the central explosion? The light from this jet was shown, by Baade, to be polarized; another piece of backing for Shklovsky's theory of synchrotron radiation as the microwave source.

Perhaps a later stage still is one in which the main source of microwave emission moves out of the galactic nucleus entirely and emerges on either side. (This seems to be true of about 3/4 of the radio galaxies studied.) In the case of NGC 5128, which radiates microwaves as intensely as M87, there are four regions of microwave emission. A pair of intense ones are found, one on either side of the central dust band; and a pair of weaker, more extended, ones are on either side of what can be seen optically. The source of microwave emission has split and moved apart toward the edge of the galactic nucleus with some of it having been hurled far beyond the nucleus in either direction. Can it be then that the dust band is not the rim of a spiral galaxy moving into a spheroidal galaxy, as was first suggested, but that it is instead the product of whatever had been going on in the catastrophe-ridden core of the galaxy? Could the dust band be a vast cloud of disintegrated star-stuff, perhaps, that happens to be hurled out in our direction?

NGC 5128 is relatively close to us (only about fifteen million light-years away), and it can be seen in some detail. If it were much farther away, the dust band and all that surrounded it would shrink until all that might be made out would be two patches of light not quite touching. The result might be interpreted as two galaxies coming together broadside, like a pair of cymbals.

But this is exactly the interpretation of the galaxies in Cygnus A. Perhaps that represents a case just like that of NGC 5128, but one that is seen more dimly because it is 700 million light-years away and not merely 15 million. If so, the explosion may be more advanced because now all the material giving rise to microwave emission has been hurled beyond the galactic nucleus, part to either side and the two sources are separated by 300,000 light-years. The same is true of other galaxies where the radio sources exist on either side of the galaxies themselves. The galaxies, nevertheless, still show signs of the catastrophe, for their optical spectra indicate ferociously high temperatures.

And perhaps the latest stage of all takes place when the radio sources grow more diffuse and dim until they reach the point where they can no longer be detected, and the galaxy is once more (as far as we can tell by radio astronomy) a normal galaxy.

Yet, while the notion of colliding galaxies slowly died and that of exploding galaxies became prominent, the evidence in favor of the latter remained almost entirely in the form of deduction from the nature of the microwave emission throughout the 1950's. The only piece of optical evidence in favor of the explosion theory was the case of the jet in M87, and that was dubious because it was emerging in only one direction, where one would expect such a phenomenon to take place in two opposite directions.

The necessary optical evidence came as the 1960's opened. In 1961, the American astronomer Clarence Roger Lynds (1928–   ) was trying to pinpoint a weak radio source listed as 3C231. The area fuzzily covered by the source included a number of galaxies in the constellation Ursa Major, of which the largest and most prominent was M81. It had been supposed that M81 was the radio source. However, as Lynds pinpointed the source more carefully, it zeroed in not on M81 but on a smaller neighboring galaxy, M82.

Certainly, M82 qualified as a peculiar galaxy, considerably more so than M81. Earlier photographs taken of it had shown it to be unusually dusty, and individual stars could not be made out within it although it is only ten million light-years from us (which is close enough to allow some stars to show up). In addition, the radio source was at its center and there were faint signs of filaments of gas or dust above and below it.

Once M82 was pinpointed as a radio source, however, its optical properties gained new interest. The American astronomer Allan Rex Sandage (1926–   ) took photographs with the 200-inch telescope, using a special red filter that let through chiefly the light associated with hot hydrogen. He reasoned that if something were going on in the center of that galaxy, something that spewed out matter, then that matter would be chiefly hydrogen and it would be seen most clearly if light from other sources were excluded.

He was right. The galaxy, M82, was clearly and visibly undergoing a vast explosion. A three-hour exposure revealed jets of hydrogen up to 1000 light-years long, bursting out of the galactic nucleus. The total mass of hydrogen being shot out was the equivalent of at least 5,000,000 average stars. From the rate at which the jets were traveling and from the distance they had covered, the explosion, as now seen from the Earth, must have been in progress for 1,500,000 years. Apparently, this is still an early stage of the process, too soon to develop the late-stage appearance of a double source on either side of the galaxy.

The light from M82 is polarized in such a way as to indicate that the galaxy has a strong magnetic field. Again, the synchrotron radiation theory is backed. (In 1965, it was discovered that synchrotron radiation was arriving from a halo about M81; perhaps as a response to the outpouring of energy from its exploding neighbor.)

Can it be that galactic explosions are comparatively common, that it is a stage that galaxies often pass through—just as many stars may pass through a nova stage? Has our own Galaxy passed through such a stage? Has our Galactic nucleus exploded? If it has, the explosion cannot have been a very large one, or a very recent one, for there is no sign of strong radio sources on either side of the Galaxy. On the other hand, such an explosion would have produced vast floods of neutrinos, which could have affected the proportions of some of the types of atoms in the Earth's crust and some scientists think the proportions now found in the crust bespeak such a neutrino flood.

Then, too, there is the continuing outstreaming of hydrogen from the center to the outskirts of the Galaxy (see page 273) and the fact that clouds of neutral hydrogen have been found at high galactic latitudes; that is, far from the plane of the Milky Way. Such high-placed clouds might also indicate the fact of a long-past explosion.

The Distant Radio Sources

But what causes the explosions?

Since the most colossal explosion we know of, on a subgalactic scale, is the supernova, it seems reasonable to suppose that a galactic explosion results from the simultaneous or nearly simultaneous explosion of a great many supernovae. This was suggested by the American astronomer Geoffrey R. Burbidge.

We can picture a large number of massive stars having reached the stage where they are on the point of a supernova explosion. If one of them explodes, its radiation may heat up neighboring stars a trifle, pushing some of them over the edge and driving them into explosions. They will, in turn, touch off others. This is most likely to happen at the galactic center, of course, where the stars are most closely spaced and where this domino effect of supernova formation can proceed with the greatest speed. (It takes time for radiation to cross the gap between stars.)

But even such multiple-supernovae supply barely enough energy. It would take all the hydrogen in ten billion stars like our own to supply the total microwave energy released by a giant radio source such as Cygnus A, even if it were supposed that all the energy from hydrogen fusion would go into microwaves with 100 percent efficiency. Such a catastrophe would involve every star in the nucleus of a medium-sized galaxy.

This wholesale slaughter of stars is a rather drastic way out of the difficulty. Is there any way we can economize on stars?

Suppose, for instance, that what is happening is the interaction of matter and antimatter (see page 253). Matter-antimatter annihilation converts all matter involved into energy, and not merely 1 percent of it as nuclear fusion does. The energy of Cygnus A could then be supplied not by ten billion supernovae, but by a mere hundred million stars (half matter and half anti-matter) undergoing mutual annihilation.

And yet this thought is not a comfortable one, either. We have no direct evidence of the existence of antimatter in huge masses. If such antimatter existed, it might explain something like the jet in M87 which might be pictured as caused by the invasion of a wedge of antimatter from outside the galaxy (or the invasion of a quantity of matter if M87 is itself antimatter). But how explain the appearance of radio sources on each side of a galaxy? The simultaneous and symmetrical invasion of antimatter clouds from both sides would seem unlikely. If one supposed the actual annihilation took place at the center and the two clouds to either side were the result of an explosion, how could the antimatter get to the center of the galaxy without first annihilating the outskirts?

Fred Hoyle points out, however, that one can go back to gravitational energy. If one begins with a mass great enough and postulates a sufficiently catastrophic collapse, one can account for the production of enormous quantites of energy. Indeed, the amount of energy produced in that fashion per unit mass can be a hundred times as great as that produced by ordinary nuclear fusion and therefore just as great as that produced by matter–antimatter annihilation.

One can picture a galactic nucleus containing stars so

crowded together that the mutual gravitational field is intense enough to overcome those factors that tend to keep them apart. A number of stars start to move together, intensifying the gravitational field, drawing other stars into the vortex until, at the end, the mass of something like a hundred million stars form a single black hole that, by continuing to swallow stars and to grow, can produce the necessary energy.

In fact, there is even the suggestion that the black hole came first; that at the time of the big bang, the pressures that formed the postulated mini-black holes also formed huge ones. Each black hole gradually accumulates an accretion disc about itself. The larger and more massive the black hole, the greater the accretion disk, the greater the gravitational field of the accretion disk, and the greater the tendency of the accretion disk to grow still further. About the largest black holes, billions upon billions of stars gathered and galaxies thus formed.

The advantage of the gravitational theory over the matter–antimatter theory is that no collision of any sort is required. A matter-galaxy can do it all on its own. So can an antimatter-galaxy.

But are we correct in turning only to that which is known? A century ago, Helmholtz tried to explain the enormous energy output of the Sun by means of forces then known and was forced to postulate a very short-lived Solar system. The radiation of the Sun and other stars could not be explained with real elegance until a new kind of energy, nuclear energy, was recognized. In the same way, does the still more massive output of certain radio sources indicate kinds of energies we have not yet recognized even today? Some astronomers cannot help wondering.

But whatever the source of the ferocious energy of the intense radio sources, its mere existence is of great potential use to those interested in the problems of cosmology and cosmogony.

Through the middle 1950's, the most distant ordinary galaxies that could be seen by optical telescopes were perhaps two billion light-years away. That is about one-sixth the distance to the edge of the observable Universe, and this is not quite far enough to be able to tell distinctly whether the Universe is hyperbolic, pulsation, or steady-state (see page 219).

Viewing the Universe by microwave, however, gives hope of a longer reach. Light sources are almost endlessly many, while microwave sources are few. In our own Galaxy there are billions of stars crowding in on telescopes and only a hundred radio sources. Outside our Galaxy, there are billions of ordinary galaxies but only thousands of radio galaxies. This means that individual microwave sources, even though very distant, are not likely to be lost in the clutter of nearby sources, and are therefore much more easily detected and studied than equally distant optical objects are.

Thus, Cygnus A, which is 700,000,000 light-years away, has so little competition, so to speak, from nearer radio sources that it is the second brightest radio star in the sky. Only Cassiopeia A is brighter. Furthermore, our radio telescopes can pick up a radio source as intense as Cygnus A at distances that would far outdo anything our best optical telescopes could attain. The microwave emission of Cygnus A could be picked up clearly at distances that would reduce its light to an indetectably dim flicker. In 1959, for instance, Minkowski detected a radio source with a red shift indicating it to be 4.5 billion light-years away, which placed it much farther than any optically-visible object known at that time. The distribution of the very distant radio sources might, therefore, help us choose a model of the Universe, where the distribution of the less-distant ordinary galaxies, detectable only by light, would fail.

As a first approximation one can assume that on the whole fainter radio galaxies are more distant than more intense ones, just as Hubble could work on the assumption that dimmer galaxies were more distant than brighter ones (see page 188). Using that assumption, the English astronomer Martin Ryle (1918–) attempted to analyze the manner in which the number of radio sources increased as their intensity diminished, much as Herschel had once done in the case of stars (see page 47).

If the steady-state theory of the Universe is correct, then the average distance between galaxies has always been what it is now. The radiation from distant radio sources, which were first produced eons ago and therefore represent a Universe eons younger than the Universe in our own neighborhood, should show those radio sources to be spread no more thickly through space than are the radio sources among our own neighboring galaxies. In that case the number of radio sources ought to increase, as their intensity diminished, according to a fixed formula.

Herschel had found that in the case of stars, the increase with dimness fell short of the formula, from which he deduced a finite Galaxy. Ryle found the reverse. The number of radio sources increased with dimness more rapidly than would be indicated by the formula. It seemed that radio sources in the far distance were more thickly spread than in our own neighborhood.

When he announced this in the mid-1950's, his analysis seemed to favor the big-bang and to indicate either a hyperbolic or a pulsating Universe. After all, in the youth of the Universe shortly after the big bang, the galaxies were spaced more closely and therefore the radio galaxies were spaced more closely, too. This means that in the far reaches of space, where the radiation reaching us was produced in the youth of the Universe, radio sources would be more numerous than they are here.

Ryle's data seemed, at that time, even to fit in with the colliding-galaxy theory of radio sources. If the galaxies were closer together in the far past (and therefore in the far distance), one would expect more frequent collisions and therefore more numerous radio galaxies.

The passing of the colliding-galaxy theory does not, however, weaken Ryle's case if the data he presents are valid. It may well be that exploding nuclei are a feature of a galaxy's youth, that a galaxy is more likely to experience the catastrophe in its first eon of existence than in its second, more likely to do so in its second than in its third, and so on. In that case, one would expect the explosions to be more numerous in the youth of the Universe, and therefore in the far reaches made out by our radio telescopes, than in the present Universe of our own neighborhood.

In fact, one need not account for the difference at all. The mere existence of the difference is sufficient. If there is any overall difference between the Universe here and the Universe near the edge, then the steady-state theory is eliminated, since it is the essence of the steady-state that there be *no* significant difference through space or time.

Of course, one can question Ryle's data. They rest on the detection and measurement of very faint radio sources and these can only be, at best, of limited accuracy and reliability. Hoyle, for instance, clung stubbornly to the steady-state theory, despite Ryle's findings, maintaining that the microwave data were not yet sufficiently sturdy to support a final decision.

But, then, quite unexpectedly, came a phenomenon that extended the astronomer's reach even more startlingly than did Ryle's distant radio sources and, moreover, moved him back into the optical portion of the spectrum, where, all things being equal, he could see more sharply.

Quasars

As explained earlier, many radio sources had been shown, as far back as 1948, to be point sources, originating from quite restricted areas of the sky. The average diameter of such radio sources is about 30 seconds of arc. This means that all the "radio stars" could be fitted comfortably into an area not much more than half the size of the full Moon.

And yet there are a few radio sources that are unusually small even by those standards. As methods for pinpointing the sources grew more refined, it seemed that a few were only 1 second of arc or less in diameter.

This is unusually compact, and it could not help but rouse the suspicion that if ordinary "radio stars" were actually radio galaxies, then very compact "radio stars" might really and literally be radio stars. Certainly, that would account for their compactness.

Among these compact radio sources were several known as 3C48, 3C147, 3C196, 3C273, and 3C286. The "3C" is short for "Third Cambridge Catalog of Radio Stars," a listing complied by Ryle and his group, while the remaining numbers represent the placing of the source on that list.

In 1960, the areas containing these compact radio sources were combed by Sandage with the 200-inch telescope, and in each case a star seemed to be the source, the first to be detected being 3C48. In the case of 3C273, the brightest of the objects, the precise position was obtained by Cyril Hazard in Australia who recorded the moment of blankout as the Moon passed before it. The stars involved had been recorded on previous photographic sweeps of the sky and had always been taken to be nothing more than faint members of our own Galaxy. Painstaking photographing, spurred by their unusual microwave emission, now showed, however, that that was *not* all there was to it. Faint nebulosities proved to be associated with some (but not all) of the objects, and 3C273 showed signs of a tiny jet of matter emerging from it. In fact, there are two radio sources in connection with 3C273, one from the star, and one from the jet.

The compact radio sources, although they looked like stars, might not be ordinary stars at all. They eventually came to be called "quasi-stellar sources," where "quasi-stellar" means "star-resembling." As the term became more and more important to astronomers, quasi-stellar radio sources became too inconvenient a mouthful, and it was shortened by Hong-Yee Chiu in 1964 to "quasar" ("*quasi*-stell*ar*"), an uneuphonious word that is now firmly embedded in astronomic terminology.

Clearly, the quasars were interesting enough to warrant investigation with the full battery of astronomic techniques and that meant spectroscopy. First Sandage, then Jesse L. Greenstein and Maarten Schmidt labored to obtain the spectra and when they accomplished that task in 1960 they found themselves with strange lines they could not identify. Furthermore, the lines in the spectra of one quasar did not match those in any other. This was puzzling, but the quasars were still accepted as objects of our own Galaxy.

In 1963, however, Schmidt returned to the spectrum of 3C273 which, with a magnitude of 12.8, is the brightest of the quasars. Six lines were present, of which four were spaced in such a way as to seem to resemble a series of hydrogen lines—except that no such series ought to exist in the place in which they were found. What, though, if those lines were located elsewhere but were found where they were because they had been displaced toward the red end of the spectrum? If so, it was a large displacement, one that indicated a recession at the velocity of over 25,000 miles per second. This seemed unbelievable and yet if such a displacement existed, the other two lines could also be identified; one represented oxygen minus two electrons, the other magnesium minus two electrons.

Schmidt and Greenstein turned to the other quasar spectra and found that the lines there could also be identified, provided huge red shifts were assumed.

Such enormous red shifts could be brought about by the general expansion of the Universe; but if the red shift was equated with distance in accordance with Hubble's law (see page 188), it turned out that the quasars could not be ordinary stars of our own Galaxy at all. They had to be among the most distant objects known—billions of light-years away.

This was a hard thing to accept, and alternate explanations for the red shift were sought. Could the quasars be very massive objects with gravitational fields so intense as to produce an enormous Einstein red shift (see page 160)? Could they, for instance, be regions of thin gas made luminous by hard radiation from numerous neutron stars within that gas, whose enormous gravitational fields would produce the red shift. At least one quasar, 3C-273B, was discovered, in 1967, to emit X-rays (about a billion times the quantity emitted by the Crab Nebula) and that would fit the many-neutron-star theory. And, indeed, Philip Morrison of M.I.T. has speculated that a quasar may be a giant neutron star itself and would be to a normal galaxy what a pulsar was to a normal star.

If all this were so, quasars might be close by. But then it was shown that strong theoretical objections could be raised to the possibility of the red shift being brought about through gravitational effects. Whatever the explanation, the red shift seemed to be due to velocity.

But then, might they not still be relatively close by and have attained huge velocities not through the expansion of the Universe but through some explosion at the center of our Galaxy (a "small bang" so to speak) as through the collision of neutron stars?

If so, all the quasars would be streaking away from the center of the Galaxy and in that case, it might be argued that some quasars would be aproaching us and would show an enormous violet shift. Some might be fairly close to us and show measurable proper motion. Instead, by the end of the 1960's about a hundred and fifty quasars had been discovered and of the spectra studied, every single one of them showed a large red shift, larger ones indeed than that of 3C273 and none showed a perceptible proper motion.

James Terrell of Los Alamos, who is prominent among those astronomers who dispute the great distance of the quasars, points out, however, that this is not decisive. If the explosion had taken place a very long time ago, then all the quasars which had rushed outward in the direction of the Sun would have passed us and would now be receding. All would be so far away (though not at super-distances) to show no proper motion. In 1968, Terrell even reported a proper motion for 3C273 that would indicate a distance of only 400,000 light-years.

At least one direct piece of evidence, independent of the red shift, has, however, been advanced in favor of the great distance of the quasars. The microwave emission of 3C273 seems rather deficient in the 21-centimeter region; there is an absorption line there. The absorption line itself shows a red shift that indicates that the hydrogen cloud which may be expected to absorb in that region must lie at a distance of forty million light-years. Between ourselves and 3C273, and at a distance of forty million light-years, there happens to be a cluster of galaxies in the constellation Virgo. It seems quite reasonable to suppose that there is a cloud of hydrogen gas associated with this cluster and that it is responsible for the 21-centimeter absorption.

In that case, 3C273, being farther from the Virgo cluster could certainly not be a member of our Galaxy. And if it is beyond the Virgo cluster, then it seems impossible not to accept it as being as far away as its red shift indicates. Then, if one quasar is very far away, and if its red shift represents this vast distance, it becomes difficult to argue that the even more extreme red shifts of the other quasars represent some other factor than distance.

If we accept the red shifts at face value, most of the more recently discovered quasars are more distant than those of the first batch. In 1967, Bolton found the quasar PKS 0237-23 to have a red shift so large as to indicate a recession at 82.4 percent the speed of light. (This represents a velocity of some 160,000 miles per second.) Such objects must be nearly nine billion light-years away, according to Sandage, and the light that reaches us from them must therefore have been emitted nearly 9 eons ago.

And yet the acceptance of the huge distances of the quasars involves astronomers in some puzzling and difficult points. If the quasars are indeed as incredibly distant as they seem to be from their red shifts, then they must be extraordinarily luminous to appear as bright as they do. They seem not only to be emitting a huge quantity of microwaves, but a huge quantity of visible light also. The quasars are anywhere from thirty to a hundred times as luminous as an entire ordinary galaxy! They might also account for appreciable fractions of the ultra-high energy cosmic-ray production in the Universe.

Yet if this is so, and if the quasars had the form and appearance of a galaxy, they ought to contain up to a hundred times as many stars as an ordinary galaxy and be up to five or six times as large in each dimension. Even at their enormous distances they ought to show up as distinct oval blotches of light in the 200-incher. The fact that they appeared to be starlike points seemed to indicate that, despite their unusual brightness, they had to be far smaller in size than ordinary galaxies.

The smallness in size was accentuated by another phenomenon, for, as early as 1963, the quasars were found to be variable in the energy they emitted, both in the visible light region and in the microwave region. Increases and decreases of as much as 3 magnitudes were recorded over the space of a few years.

For radiation to vary so markedly in so short a time, a body must be small. Small variations might result from brightenings and dimmings in restricted regions of a body, but large variations must involve the body as a whole. If the body is involved as a whole, some effect must make itself felt across the full width of the body within the time of variation. But no effect can travel faster than light, so that if a quasar varies markedly over a period of a few years, it cannot be more than a light-year or so in diameter. (Actually, some calculations have shown that quasars may be as little as a light-week—500 billion miles—in diameter.)

This combination of tiny volume and great luminosity poses such problems to the astronomer that there is a constant tendency to find some way of accepting the quasars as nearby bodies after all. If the quasars were nearby, as Terrell maintains, they would not have to be exceedingly luminous to be as bright as they appear; they would be no more luminous than one would expect a body a light-year across to be. But if we assume that the quasars are really vastly distant, then we are faced with a body that is only a light-year across and yet is up to a hundred times as luminous as an ordinary galaxy that is a hundred thousand light-years across. How can that be explained?

A number of possibilities have been offered. One such is that of the "gravitational lens." It is supposed that one galaxy may be directly in front of another as seen from Earth. The light from the galaxy in the rear is bent around the first by gravitational effects and that light is then concentrated in our direction. It would work like the lens of a telescope and would increase the apparent brightness of the rear galaxy enough to make the two together appear to us like a super luminous quasar.

More likely perhaps is some sort of implosion, some catastrophe on the galactic scale that is like the supernova is on the stellar scale. Could it be that on a galactic scale, ordinary exploding galaxies are "novae" that radiate much energy but retain their substance and general galaxylike form, while some particularly enormous catastrophes are like "supernovae" so that much of the galactic substance is blown away while what remains contracts catastrophically into a quasar? (And, indeed, the light received from quasars is similar in some respects to that received from white dwarfs!)

In that case quasars might even be the galactic equivalent of neutron stars, as Morrison suggests, even though the red shift would still not be due to its gravitational field but to its velocity.

If anything like this is the proper interpretation of a quasar, it can only be a short-lived object. It cannot radiate such immense quantities of radiation for very long. Some calculations indicate that it can exist as a quasar for only a million years or so. In that case, the quasars we see only became quasars a short time ago, cosmically speaking, and there must be a number of objects that were once quasars but are quasars no longer.

Sandage, in 1965, announced the discovery of objects that may indeed be aged quasars. They seemed like ordinary bluish stars, but they possessed huge red shifts as quasars did. They were as distant, as luminous, as small as quasars, but they lacked the microwave emission. He called them "blue stellar objects" which can be abbreviated as BSO's.

The BSO's seem to be more numerous than quasars. A 1967 estimate places the total number of BSO's within reach of our telescopes at 100,000. If they develop out of quasars, then they are fifty times as numerous because they endure in the BSO form fifty times as long—say 50,000,000 years. Still older quasars must dim to the point where they can be detected neither by microwave emission nor by light emission. What shape they may then take or how they may then be recognized is as yet unknown.

The mere existence of quasars and BSO's has been considered a heavy blow at the steady-state theory. If all of them are very far away and were therefore formed many eons ago, and if none can be detected in our own neighborhood, it seems that whatever processes formed them are not operative now (although there may be long-dead quasars in our neighborhood that we have not yet learned to recognize). This, in turn, means that the Universe was different in important ways, eons ago.

If the big bang took place some 15 eons ago, this makes sense. The Universe was smaller, hotter, younger, and fuller eons ago than it is today, and it is not at all surprising that catastrophic events such as quasar-formation should be common then while in today's larger, cooler, older, emptier Universe they do not occur.

Arguing against this is the fact that, despite everything, the decade of the 1960's has not indisputably proven that quasars are extremely distant and the thought that they might be only moderately distant remains alive. Nor do they seem as unique as was thought at first.

Back in 1943, Carl Seyfert, while still a graduate student in astronomy, detected a peculiar galaxy, which has since been recognized as one of a group now called "Seyfert galaxies." They may possibly make up 1 percent of all known galaxies though actually by the end of the 1960's only a dozen examples were known.

In most respects, Seyfert galaxies are normal and have only moderate red-shifts. The nuclei, however, are very compact, very bright, and seem unusually hot and active—rather quasar-like, in fact. They show variations in radiation that imply the radio-emitting centers at their core are no larger than quasars are thought to be.

The strongly active center would be visible at greater distances than the outer layers of the Seyfert galaxy would be, and if such a galaxy were far enough, all we would see by either optical or radio telescopes would be the core. We would then consider it a quasar. The very distant quasars may simply be very large Seyfert galaxies.

If this is so, quasars may be forming relatively close to us and it may be a phenomenon not restricted to great distances and therefore not indicating so definitely that the Universe was quite different many billions of years ago.

If the quasars have weakened as support for the big bang theory, however, another type of observation has taken over that support, and most dramatically.

If the Universe had indeed begun with a big bang that explosion must have emitted an enormous flood of very energetic radiation—x-rays and gamma rays. If we probe out into the very far distance, billions of light-years away, we would be reaching, in time, billions of years ago and we might be able to see this fire-ball.

What would the fire-ball look like? Since it would be very far away, up to 15,000,000 light-years away, it would suffer an enormous red-shift and all that very energetic radiation would appear in the radio-wave region.

In fact, we wouldn't be able to see the fire-ball itself since, for a period of time after the big bang, the expanding but still-small Universe would be dense with radiation and matter would not have formed. It would be only after a substantial period of time that matter would have become predominant, that energy would thin out and allow the stars and galaxies to form, leaving space transparent.

Sure enough, beyond the farthest quasars we can see there seems to be nothing even though, if there were quasars there, we could see them easily. Perhaps we have penetrated to the opaque, energy-thick space of the infant Universe.

Another way of looking at it is that as the Universe expanded it cooled down, and it should now be at the average temperature fairly close to absolute zero. This means that the Universe generally should produce the kind of radiation that a system at this temperature should produce; that is, long-wave radio waves appearing at equal intensity from every direction. It would be a kind of "isotropic radio-wave background."

George Gamow pointed out the likelihood of the existence of this radio-wave background in the 1940's. In 1964, Robert H. Dicke of Princeton University (1916–   ) revived the notion.

The search for the background radiation began at once, and in 1965, Arno A. Penzias (1933–   ) and Robert W. Wilson (1936–   ) of Bell Telephone Laboratories reported that after all microwave emission was accounted for, there was still a faint general background of radiation. Men had heard the echo of the big bang still resounding through the Universe.

To be sure, judging from the nature of the radiation, the present temperature of it would seem to be $3°$ K, or three degrees above absolute zero, instead of the $10°$ K that was predicted by Dicke. The discrepancy seems to indicate that the original temperature of the big bang was less than had seemed likely. This could mean that the Universe at birth was not pure hydrogen but contained a considerable admixture of helium.

* * * * *

And so, as I promised at the start, we have followed the quest of mankind out toward the endlessly receding horizon. We began with man's narrow vision of a patch of flat Earth and have paused now at the point where man pictures a Universe 26,000,-000,000 light-years in diameter.

Nor need we feel that we have now plumbed the Universe to the utmost. Astronomy has been advancing at an ever-accelerating pace for four centuries now, and there are no signs, as yet, of any leveling off. More has been learned about the Universe in the last generation than in all man's history before; what, then, may lie ahead in the next generation?

If it is exciting to probe the unknown and shed light on what was dark before, then more and more excitement surely lies ahead of us.

Suggested Further Reading

Abell, George O., *Exploration of the Universe*, 3rd edition, Holt, Rinehart & Winston, New York (1974).

Alfven, Hannes, *Worlds-Antiworlds*, Freeman, San Francisco (1966).

Bonnor, William, *The Mystery of the Expanding Universe*, The Macmillan Co., New York (1964).

Burbidge, Geoffrey and Burbidge, Margaret, *Quasi-Stellar Objects*, Freeman, San Francisco (1967).

The Cambridge Encyclopedia of Astronomy, Crown, New York (1977).

Glasstone, Samuel, *Sourcebook on the Space Sciences,* Van Nostrand Co., Inc., Princeton (1965).

Hoyle, Fred, *Galaxies, Nuclei and Quasars*, Harper & Row, Publishers, New York (1965).

Ley, Willy, *Watchers of the Skies*, Viking, New York (1966).

McLoughlin, Dean B., *Introduction to Astronomy*, Houghton Mifflin Co., Boston (1961).

Motz, Lloyd, *The Universe: Its Beginning and End*, Scribner, New York (1975).

Pannekoek, A., *A History of Astronomy*, Interscience Publishers, New York (1961).

Sciama, D. W., *The Physical Foundations of General Relativity*, Doubleday, New York (1969).

Shklovskii, I. S. and Sagan, Carl, *Intelligent Life in the Universe,* Holden-Day, San Francisco (1966).

Struve, Otto, and Zebergs, Velta, *Astronomy of the 20th Century,* The Macmillan Co., New York (1962).

Sullivan, Walter, *Black Holes*, Doubleday, New York (1979).

Index

320